Building Owner's and Manager's Guide: Optimizing Facility Performance

BUILDING OWNER'S AND MANAGER'S GUIDE: OPTIMIZING FACILITY PERFORMANCE

By
ROBERT S. CURL, P.E.

Published by
THE FAIRMONT PRESS, INC.
700 Indian Trail
Lilburn, GA 30047

Library of Congress Cataloging-in-Publication Data

Curl, Robert S.
Building owner's and manager's guide : optimizing facility performance / by
Robert S. Curl.
   p. cm.
   Includes index.
   ISBN 0-88173-290-7
   1. Building management. I. Title.
TX955.C87   1998   658.2--dc21   98-38503
                                                    CIP

Building owner's and manager's guide : optimizing facility performance / by Robert S. Curl.

Published by The Fairmont Press, Inc.
700 Indian Trail
Lilburn, GA 30047

Printed in the United States of America

10 9 8 7 6 5 4 3 2 1

ISBN 0-88173-290-7 FP

ISBN 0-13-083831-4 PH

While every effort is made to provide dependable information, the publisher, authors, and
editors cannot be held responsible for any errors or omissions.

Distributed by Prentice Hall PTR
Prentice-Hall, Inc.
A Simon & Schuster Company
Upper Saddle River, NJ 07458

Prentice-Hall International (UK) Limited, London
Prentice-Hall of Australia Pty. Limited, Sydney
Prentice-Hall Canada Inc., Toronto
Prentice-Hall Hispanoamericana, S.A., Mexico
Prentice-Hall of India Private Limited, New Delhi
Prentice-Hall of Japan, Inc., Tokyo
Simon & Schuster Asia Pte. Ltd., Singapore
Editora Prentice-Hall do Brasil, Ltda., Rio de Janeiro

Dedicated To
Ruth Curl
For Support and Understanding

and

Judy Geddis
Whose Flying Fingers Put This Together

CONTENTS

Circulation; Water Treatment; Lighting; Building Repairs; Workplace Arrangement

ing Room Control; Vocational School Control; Flexible Ducts; Toilet and Kitchen Exhaust; Air Conditioning Unit Drain Pans; VAV Systems Under Roof; Heat Pumps; Computer Room Cooling; Reheat Systems; Cooling Unoccupied Space; Humidifiers; Enthalpy Control; Water Meters; Piping Flexible Connections; Engine Generator Piping; Electrical Grounding; Pipe Insulation Sealing; Low Level Air Supply; VAV Fan Box; Supply Air Patterns; Steam Pressure Increase for Turbines; Duct to Outlet Connections; Direct Fired Gas Make-up Units; Lighting Arrangement; High Ceiling Church Cooling; Atrium Air Conditioning; Lobby Air Conditioning; Corridor Temperature; Window Radiant Heating; Fountains; Elevator Machine Room; Wall Air Leaks; Outside Air Supply to Ceiling R.A.; Air Supply to Toilets; Coil Freeze-up; Cogeneration; Well Water; Piping Labels and Color; Vibration Isolation; Noise Pollution; Piping Expansion Joints; Duct Flexible Connections; Isolation for Electrolysis; Damper Motor Operation; Indirect Air Supply; Access to Air Conditioning Equipment; Refrigerant Piping Internal Condition; Filters; Roof Duct Installation; Laboratories; Libraries; Museums; Bakeries; Laundries; Athletic Facilities; Indoor Swimming Pools; Auditoriums; Theaters; Outside Air Intake Below Ground; Cooling Towers Below Ground; Residences; Boiler Economizer; Smoking Room; Smoke Removal Systems

Understandable Screen; Sensor Location; Unoccupied Space Control; Maintenance Information; Cabinet Unit Control; Return Air Fans; Time Clocks; Step Start-up; Kitchen Hood Operation; Portable Thermometer; Burner Oil Supply; Security Alarms

INTRODUCTION

The three most important factors to provide for building occupants is comfort, health, and operating costs.

Up to now, most of the interest has been on energy cost reduction to reach a BOMA average of about $1.65 per square foot per year.

Health is becoming a factor in buildings with "Sick Building Syndrome."

One operating cost least considered is employee comfort. The average cost of $130 per person per square foot per year greatly outweighs the energy savings.

Chapter 1

General Conditions— All Buildings

This guide is prepared for people who manage buildings and are required to keep the occupants, tenants, and owners happy, and to pay their bills without complaint.

In general, this requires that the building be comfortable from the standpoint of temperature, humidity, and air circulation; have good lighting and pleasant surroundings; and be provided with the necessary services such as sanitation waste removal, access, and parking space. In addition, security is becoming more and more of interest and the expansion of communication skills has become so great that there must be some protection against loss of information and theft of confidentiality.

In this manual, we will be discussing primarily comfort, health, operating costs, and legal liability. Since computers and automation cannot solve all problems, we are presenting some simple ways of checking on these conditions with minimum cost for instrumentation and time.

Atmospheric comfort is a function of temperature and humidity. The normal body seated at rest gives off about 400 Btu per hour of heat, about 60% of which is radiated to the surrounding atmosphere, and 40% is moisture evaporated from the skin. Consequently, there should be air circulation around the body so that this heat and humidity can be removed. There should be a temperature difference below the body temperature (normal at 98.6° F), and the humidity should be such that moisture can evaporate from the body, to avoid excessive perspiration. If you are walking fast or working hard, your body heat loss may get up to 1,700 Btu per hour, which will require more air motion and probably lower temperature and humidity.

Other things may effect your body condition such as stress, anxiety, pressure to produce, poor general health, and sometimes simply psychological influences.

I have worked with architects and interior decorators who have determined that certain colors, pictures, and decorations can help make a person feel more comfortable in a space. Of course, towards the end of the day, fatigue and stress may make a change in the comfort of a building's occupants.

Certain materials in the atmosphere can also cause a deterioration in health such as sneezing, coughing and general discomfort. These can be caused by mold, mildew, dust, certain gases, and poor air circulation. This is particularly true with some of the new machines that are used in offices, such as copying machines which give off carbon dust and ozone, and the heat and reflection from computers, and even residual odors from cleaning compounds. Many people are allergic to certain fragrances which are supposed to make the air fresher, but sometimes cause difficulty in the respiratory system.

Many of the problems you encounter were built-in because of the original design of the building. Chapters 12 to 21 cover items and conditions that the architect and the engineer should have taken care of when they designed the building and for which the contractor should have made provisions. If these items had been considered originally, many problems would have been forestalled and your maintenance costs reduced considerably.

If you own and operate a building, regardless of its age, many of the items discussed under Original Design may provide for modifications at minimum cost to improve both health conditions and the cost of operation.

Although a lot of the information may indicate problems during the cooling cycle of air conditioning and especially in high-humidity areas, the author has had 40 years' experience designing systems in cold areas with temperatures as low as 100° below zero.

In one building we designed in an area which experienced 30° below zero temperature and 50 mph wind, we discovered that there was sufficient outside air pressure to blow snow through the keyhole into the living room. We had to prepare the walls in such a manner that not only did they insulate, but they prevented high pressure air from entering the space.

CHAPTER 2

GENERAL CONDITIONS — EXISTING BUILDINGS

In general, the first part of this book discusses the operation, corrections, and procedures for existing buildings. In the latter part of the book is a manual on what to look for in the design of a new building, and what should be presented to the architect and engineer to make sure that they include this information in the drawings and specifications.

These recommendations and suggestions are based on 60 years' experience of a design engineer on mechanical and electrical services for a billion dollars worth of buildings, on field tests and solutions to problems in over 200 existing buildings, and on the experience of operating two office buildings of his own and the problems encountered therewith.

As you will note, field examination may not require very expensive equipment, and should be conducted before you call in a consultant or an environmental tester. You may be able to solve most of the problems yourself at minimum cost. This is also true in replacing operating equipment which may or may not be in good condition. The operating condition may be such that even new equipment will not provide the efficiency expected. Old equipment, if properly maintained, may give you years of service without requiring replacement.

We have experience in testing a 32-year-old building, where the air conditioning units were still in reasonably good shape, and needed only patching of the rust holes in the cabinet. The piping had rusted on the outside due to poor installation of the insulation. The saving on this one building was more than 3/4 of a million dollars.

CHAPTER 3

COMFORT CONDITIONS

Keeping the occupants comfortable and healthy so they can produce at the best rate makes it necessary to have their surroundings and their feelings in a comfortable state. This includes not only temperature and humidity, but also air circulation, the condition of any contaminates in the air, objectionable gases and odors, noise, and the appearance of the surroundings.

Have you ever had this complaint (see Figure 3-1)? Do you know how to fix it?

COSTLY THINGS HAPPEN
ON THE WAY INTO THE OFFICE...

- Cool blast of air hits you in the face walking in the front door
- Elevator lobby chilly
- Cool draft on back entering elevator
- Floor corridor cool then warm
- Step in office—warm
- Look down at thermostat, set at 72°
- Turn thermostat down to 68°
- Work for an hour, start getting cold
- Turn thermostat back up to 72°
- Sun comes out, your back gets hot
- Turn thermostat back down to 68°
- Secretary comes in wearing sweater, saying "It's freezing... "
- Call the building manager... "Can you fix this '!@-#S-%!!' air conditioning?"
- Building manager calls maintenance
- Maintenance comes over, they can't find anything wrong
- Tenant is still angry

Figure 3-1.

In building operating costs in past years, we have been stressing energy conservation as a major saving program. Today we are talking about dealing with "sick building syndrome" as another method of helping our workers stay healthy.

Most of us have forgotten that the average cost of an employee working in a building is $130.00 per square foot per year. If we can do something to increase the productivity of the employees, we have saved considerably more.

Although it is a custom to cram as many people as possible in a space because of the cost of rental, you might consider the fact that the conditions surrounding the work area have a lot to do with a person's capability and the amount of work that they can produce over the normal work day.

Rarely considered are the color and decoration of those spaces where a person working at a computer screen might rest their eyes by looking at a distance. The view does not necessary have to be beautiful, but can be pleasant if the walls are light colored and there are some paintings or other decorations. We experienced this about 40 years ago when an architect who was a disciple of Frank Lloyd Wright designed an office for a telephone company which had about 50 women in it, and he used very light colors with pleasant decorations. The employees appeared to be happy. The only problem they had in this particular space was that the air near the floor was cold and their feet were uncomfortable. This condition will be explained under the section of "Sick Building Syndrome."

Arrangement of furniture also is very important. To be convenient, seating devices should be comfortable and the work space should not be so confined that workers feel like they are sitting in the bottom of a manhole. Soundproofing material, of course, was used in a lot of partitions but the height of the partitions should be such that it does not interfere with the air circulation in the room and cause physical discomfort this way.

If you are concerned about lost time, make provisions that water coolers, central office machines like big copiers, and toilet rooms do not require an extensive trip from the work space to their location. This is particularly true in those office suites which are not completely occupied and some of the offices may have been temporarily vacated. It might be a good idea to concentrate the occupied spaces in one area with the copying machine within a short walking distance (see Figure 3-2).

If you have windows exposed to the sun, it is essential that these be shaded, as the sun penetration will cause discomfort due to radiant heat, even though the glass may be solar type or even double glass or mirrored. The surface temperature may get up to 145° with the sun shining on it which will re-radiate heat out to the person's body who is within 3 to 5 feet of that spot. One of the methods of reducing the radiant heat is to sit with the side of your body facing the window rather than your back or your front, since a cross-sectional surface of your body offers less area for heat absorption.

Even though your architect and engineer laid out the lighting in the ceiling in an even pattern because they didn't know what you were going to do with the space, reflection of that light onto the work space may cause difficulty and should be investigated. We have rearranged lighting to take care of work spaces in an office, providing the same amount of light in the proper spot and reducing the total wattage from 3 watts per square foot of floor space to 1-1/2 watts per square foot, saving quite a bit in operating costs.

If you have a break or dining area, provide extra fresh, outside air to change the atmospheric condition and remove any odors which may remain in this area due to cooking or other foods which may produce contaminates.

Temperatures and humidities in entrance lobbies, elevator lobbies, corridors, and toilet rooms should be controlled as indicated under "Building Operation Conditions."

Don't forget that board and meeting rooms need to be air conditioned separately and that the construction should be such that they are comparatively soundproof.

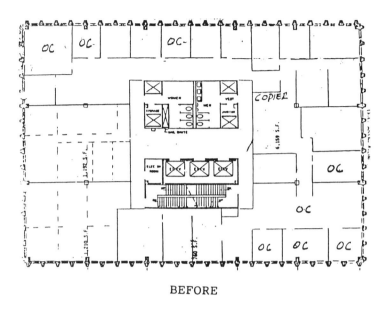

BEFORE

LOWER AIR CONDITIONING SUPPLY

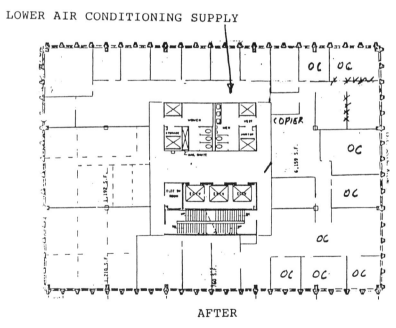

AFTER

Figure 3-2. Rearrangement of partially occupied office space to reduce walking distance.

With regard to individual office arrangements, construction should be such that there is reasonable prevention of sound transmission. Some years ago, we air conditioned an office building of a major steel company and the chairman of the board had a corner office with an individual air conditioner in his room. He complained bitterly about the noise in his office. When we visited the office, we discovered that the noise was coming from the adjoining secretarial pool and was the noise of typing. Since the partition was not that soundproof, we modified the system by putting a paper clip in the fan of the air conditioner to create a slightly different noise than what he was hearing through the partition, and the executive appeared to be satisfied.

There have been some efforts and some designs recently to provide individual air conditioning for each work space, especially those enclosed on two or three sides, by supplying the air from underneath, on the floor, and by having individual lighting in each of these spaces. It is suggested that before considering a floor supply, you investigate a similar installation; we have experienced considerable objection, from women especially, when air is supplied over feet, ankles, and legs which have minimum clothing protection. One woman put an electric heater under her desk to keep her feet warm.

If you are using carpeting, you will notice that in the areas with dark carpet the ceiling surrounding an air outlet will get dirtier faster than areas which have light carpet. Have your cleaning people vacuum the ceiling area around the outlet since this is nothing more than carpet fibers being induced up into the air outlet by its normal operation.

Check with your cleaning and janitorial service to find out if they are using any objectionable cleaning compounds which may remain in the space and be picked up in the morning and affect your personnel. This is also discussed under "Sick Building Syndrome."

Arrange the office so that workers can get out of the space easily at the end of the work day, since this is the period when they are under the most stress to get home to do other functions. Simple, straight corridors and access to elevators will help.

In toilet rooms, it is recommended that the light level be just sufficient to use the space and that the temperature be slightly above the surrounding areas to prevent employees from spending too much time. One building owner stated that he wanted the light level such that one could not read a newspaper, discouraging employees from wasting time.

CHAPTER 4

PROCEDURE FOR SOLVING COMPLAINTS

TEMPERATURE

If you have a computer management system, see what the temperature is in the complaint area. Ask someone to look at the thermostat and read the temperature and the thermostat setting. If more than 2° different from your reading, the thermostat is incorrect and should be reset, calibrated, and a label put on it so that the user can tell what the correct temperature is.

If you do not have a computer, get a thermometer with an extension bulb and swing it in the air near the complainant's location and at the thermostat (see Figure 4-1). Swing the thermometer until the temperature settles down within at least 1/2 a degree. If the thermostat is OK, is sun shining on it, is there heat near it such as a table lamp, is there a blast of cool or warm air blowing on it? If the temperature is between 72° and 75°, most people are comfortable.

Find out what the supply air temperature to the room is by putting the thermometer bulb or sensing device up in an air outlet or in the discharge duct from the air conditioner. In summer on cooling, the temperature should be between 55° and 60° in humid areas, and 50° to 55° in dry areas. In winter, 95° to 105° is preferable, except 120° can be used if properly distributed.

First, stand in the doorway coming into the building holding a handkerchief by one corner and let it hang down vertically. If it floats more than about 45°, you have air coming into the building. If it bends towards the inside of the building, at the rate of about 400 feet per minute, average costs to you are $4.00 per hour for the 8- to 10-hour work day. If that handkerchief is floating out at ap-

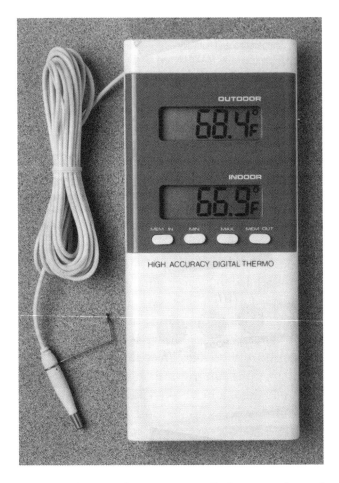

Figure 4-1. Simple test thermometer. Indoor-outdoor thermometer. Use outdoor bulb to swing in air for faster readings.

proximately a 90° angle, you are getting 1,000 feet per minute air velocity, costing you about $8 per hour, and if it is up fluttering in the breeze, you've got real trouble.

The first thing you do is look to see where that suction is coming from. The first place to go is in front of the elevators. Open the door to see if the suction is going up the elevator shaft. If it is, you'll have to check the elevator penthouse which we will discuss a little later. In the meantime, check the temperatures in the lobby, at the elevator lobby, and in the corridors of the various floors. We have found that running temperatures too low in elevator lobbies and corridors gives you a false feeling of cooling before entering your suite. If it is a 90° day, your regular lobby should be somewhere around 78°, the entrance to the elevators about 76°, the corridors between 75° and 76°, and let the tenant set his thermostat where he wants it.

In winter the opposite is true because it is cold outside and the wind is sucked into the building. You want to keep the lobby somewhere around 68° to 70°, the elevator lobby about 70°, and the corridors about 72° so that when your tenant goes into his office and he sets his thermostat at 74°, he will be warm.

If you have a receptionist or a guard in the lobby, don't forget that they have to be comfortable, so you may need an auxiliary heating or cooling unit in that particular space. Also, if the lobby is used for meeting purposes, you may have to raise the temperature or lower the temperature, depending on the season, to a comfort level of roughly between 74° and 78°.

It is much better to have the building under slight positive pressure to eliminate that blast of air coming in the front door. This will also help prevent moist air in summer and cold air in winter from being drawn into the building around windows and through walls even though they are sealed. We have experienced air leakage in all-glass buildings, air leakage through walls sufficient to make the insulation black with mold, moisture coming through walls sufficient to loosen wall covering, and moisture through roofs discoloring ceiling panels.

Unless the walls are of substantial mass such as masonry, you will get a variation in surface temperature of the outside walls depending on the weather, which will radiate heat to your body when the sun shines in summer, and will withdraw heat from your body if it is very cold on the outside. The use of solar or reflective glass does reduce some of the heat, but not all of it. You will see in the report on windows that there is a high temperature on the glass when the sun shines on it.

If you are concerned about your electric bill, and it exceeds the BOMA rate for your area, have someone look at your demand meter during the day, take a look at it on a weekday evening, look at it on Saturday and on Sunday, and see if the equipment is operating all the time. We did this on a 56,000-square-foot building which had a demand of 360 kW and had 13 air conditioning units in the building varying from 10 to 15 tons. Since I live not too far from this building, I went there during the day, at night, on Saturdays and on Sundays. I found that the air conditioning system was running all the time, even though they had an energy management system which was not functioning because the maintenance man did not know how to run it. We put in a simple system that would shut the air conditioning units off at night, turn them on in the morning, shut them off on Saturdays and Sundays, but turn them back on for two hours each day to eliminate the possibility of mold and mildew generation. This system started the motors up in steps with 3- to 5-minute intervals instead of all at one time, and we reduced the demand from 360 kW to 247 kW, saving them $5,000 a month. The first month's savings paid for the cost of the control system. In this same building, these units were air cooled and the condenser air suction and discharge went out the same louver on the outside of the building. Unfortunately, the original contractor did not make provisions to separate the air patterns with a high velocity discharge so that part of the air recirculated; on a 90° day the entering air was 107° which reduced the efficiency of the unit down to 65%. A contractor was trying to sell them a new unit with a high energy efficiency ratio (EER), which would have done them no good at all since they hadn't corrected the original poor condition.

The next step is to go up to the elevator penthouse if there is suction up the elevator shaft and find out if it is an electric elevator, has an exhaust fan and an air intake. We have found in both high-temperature and high-humidity areas, that it is cheaper to put in a small self-contained 4-ton air conditioner for a three-elevator group, let it recalculate, and shut off and seal up the exhaust fan and the air intake. In one case, we were able to stop 4,000 cubic feet of air per minute from going up the shaft and costing several thousand dollars a year in electric operating costs. The dampers on the air intake and the discharge on fans do not always close tightly, which is another cause of this suction.

Another thing to look for, if you have exhaust fans on the roof, is back-draft dampers on these fans; when the fans shut off, the dampers close off (see Figure 4-2). This is particularly true with toilet exhaust fans which should be shut off at night time when the building is not being used to prevent outside air from coming in the building and causing difficulties.

If you have parts of the building where the tenant requires air conditioning 24 hours a day, 7 days a week like computer rooms,

Figure 4-2. Back draft dampers on fan discharge.

it is advisable to provide separate equipment for that purpose. One of the recommendations you will find for the initial design of buildings using heat pump units is to provide a means for shutting down all of the system except equipment supplying those computer areas and operate with a minimum circulating pump on the condenser.

If your building is of any size, and you have continuous maintenance, there are some recommendations from the standpoint of computer management and how to adapt it to the normal use of the building with minimum operator training and minimum time to examine the conditions in the building.

HUMIDITY

General humidity should be between 40% and 60% for comfort. In winter in the north, 30% may be difficult to reach. In summer in the south, 60% may be similarly a problem. If you don't have a humidity meter (or a sling psychrometer), put a shoelace over your thermometer bulb or sensor and wet it. Swing it around until the temperature level stops. This is the wet bulb temperature. The difference between the temperature when dry, the dry bulb temperature, and the wet bulb temperature indicates about 4% relative humidity per degree below 100% relative humidity.

AIR MOTION

If you have an air meter such as a fan type anemometer, hold it up near the complainant. If there is little or no motion, a person cannot lose heat and moisture and will be uncomfortable. If you do not have an air meter, cut a facial tissue lengthwise in half and hold it up by the top edge. If there is air motion, it will move or flutter.

STRATIFICATION

Check for air motion under air outlets near the floor and at the farthest occupied space from the air outlet to see if cooling air is dumping to the floor and creating cold feet or is not spreading to the entire occupied area. With heating, do the same except check air flow at the ceiling instead of the floor. Low air flow outlets dump cold air to the floor on cooling and keep heat at the ceiling on heating (see Figure 4-3)

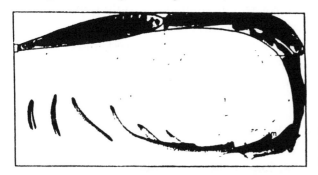

Grille supply vanes tilted slightly upward.

Normal ceiling air outlet pattern. Air induced upward in center.

Low volume air supply air falls to floor. No circulation at breathing level.

Figure 4-3. Air flow patterns.

ODORS

Lean over the copying machine while it is operating and take a deep breath. You will smell Ozone. Don't smell too long; it may make you cough and dry your throat. Bend down and smell carpeting, furniture padding, clothing, draperies, and even old books for unusual or mildew odors.

Look for black spots on air outlets, which are usually caused by mold. Dirt on ceilings near outlets is usually carpet lint and dust. If you have dark carpets in an area, the ceiling may appear dirtier than where light colored carpeting is used.

If fragrances or chemical additives are used, move allergic persons to an area of low concentration of odors and see if they feel better. Some people are allergic to fragrances.

PHYSICAL CONDITION

If none of these actions makes a complainant feel better, ask them about their physical condition, whether there is any illness that may cause them to feel uncomfortable at different temperatures. Also, some people object to air motion over their bodies. Stress is also a problem which may make a person uncomfortable because of the pressures of the job or the fact that they may be working in a small, enclosed area which feels like a cubbyhole and would like to be in an open space where they can see somebody else.

SURROUNDINGS

The physical and visual comfort of walls, furniture, flooring, and general surroundings has a lot to do with personal comfort. So does having the types of chairs or seating equipment, tables, and computer stands which provide comfort. The distance a person has to go between their location and their supervisor's, or even to office machines has some effect on fatigue, especially if it means going to a different floor even though there are elevators.

NOISE

Check for the noise level in the area and see if objectionable sounds are coming from machinery such as air conditioning fans or other sources, and try to provide for means of reducing these noise levels, as they are quite irritating to many people. If it is fan noise in an air conditioning unit, have it checked for speed, cleanliness, and whether it is vibrating. If it is office equipment that is noisy, it may be a good idea to enclose it in a separate area with a door, and supply and exhaust that area separately. If you are experiencing vibration from machinery, check to see if it is possible to isolate this equipment with heavy foundations and with vibration isolators.

PSYCHOLOGICAL

Sometimes just showing a person an accurate thermometer and humidity indicator, preferably of the digital type (and even better if the light is flashing on it) may satisfy them. If there is a real fear of serious contaminates in the air, have someone come in and run analytical tests on the chemicals in the air. This can be performed with a plunger-type instrument with glass tubing which discolors under concentration of certain chemicals. We have done this in the past and even though we did not find any of these chemicals in existence, just the fact that we ran the tests seemed to satisfy the tenants.

BUILDING OPERATION

After you have done all of these operations and still have not satisfied the party, you still may have a negative pressure in the building which is bringing particles in from the outside which cannot be seen or even collected. So it is best to change the operation of your air handling system so the building is under slight positive pressure both summer and winter.

FIELD MEASUREMENTS

If you don't want to call in an expert, you can do a lot of these tests yourself. Depending on how far you want to go, you do not need all of these instruments. Many have 2% to 5% accuracy which should be enough for ordinary purposes.

Cheaper equipment may be slower to reach final reading, but still provides the information needed. Instruments with relative cost are shown in Figures 4-4 to 4-10.

If you wish to make better measurements of quality of equipment operation and results, you will need some instruments:

- To measure temperature and humidity, use a sling psychrometer which reads dry bulb and wet bulb temperatures. You have to wet the sock on the wet bulb thermometer. The dry bulb is the actual temperature. The wet bulb is the temperature at 100% relative humidity. Using these two readings and a chart, you can determine the actual relative humidity and dew point. You have to swing the unit in the air until the reading stops changing. Digital equipment is also available which is direct reading but is slow and requires calibrating. This and other individual thermometers are shown in Figure 4.4.

- To measure air flow, the fan type thermometer is best, especially at minimum air motion, so you can watch the blades turn. This is better than using a handkerchief or facial tissue.

    Most units make one complete needle rotation indicating 100-foot velocity. If you are measuring a grille, the air velocity times the area of the outlet is square feet, which will give you cfm (cubic feet per minute). This is the usual designation for air volume.

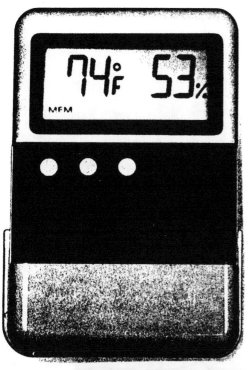

Left: Hygrometer. Reads relative humidity direct. Takes a reading in about 1/2 hour.

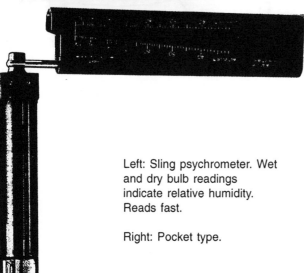

Left: Sling psychrometer. Wet and dry bulb readings indicate relative humidity. Reads fast.

Right: Pocket type.

Figure 4-4. Thermometers and humidity indicators.

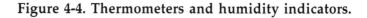

Velometer. Hand held for grille velocity.

Anemometer. Hand held for accurate air motion.

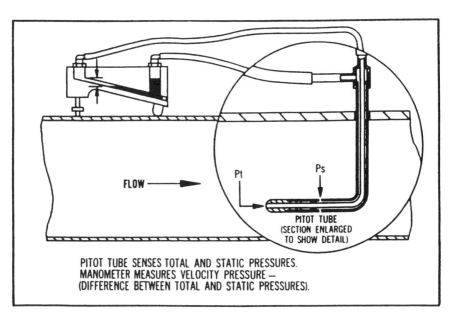

Figure 4-5. Air flow indicators.

580	520	680	550	1000	550	480	460	620	380	850	550
640	660	700	750	300	660	320	580	450	400	700	1000
580	400	630	640	380	520	530	400	500	500	550	700
460	560	920	1000	430	620	380	320	360	350	750	700
300	320	510	760	680	680	450	450	800	700	900	650
380	520	1000	950	710	950	800	800	900	900	800	650
480	550	1050	600	780	880	870	750	780	730	600	850
900	520	1050	680	650	740	730	780	970	1000	550	850
650	630	630	600	380	540	840	630	750	650	650	1000
700	1000	600	1000	320	850	460	660	670	550	450	1000

Coil air velocity distribution, 10 ft. by 12 ft. area.

Figure 4-6. Coil air velocity test.

Other analog and digital instruments such as velometers, hot wire anemometers and Pitot tubes, are good but measure in such a small area that they require many readings which are averaged. Even with the fan anemometer, several readings may be required (see Figure 4-6).

• Pressure—Depending on the level of pressure, for instance in inches of water height, or pounds per square inch, different instruments are used.

In nearly every case, valves or other shut-off devices should be installed on the connection so the gauge can be removed for calibration without disturbing the attached liquid or gas. For low pressure, liquid-filled inclined or U tubes or diaphragm gauges are used; for higher pressures, Bourdon tubes moving needles on a dial are most popular. Gauges on

Low pressure gauges for filter condition for Pitot tube velocity measurements.

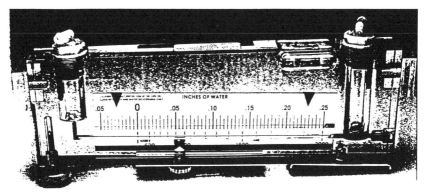

High pressure 30" vacuum to 1000 psi.

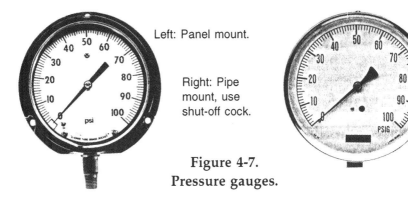

Left: Panel mount.

Right: Pipe mount, use shut-off cock.

**Figure 4-7.
Pressure gauges.**

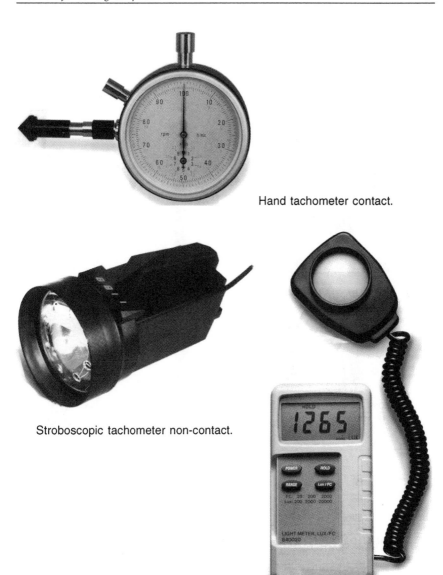

Hand tachometer contact.

Stroboscopic tachometer non-contact.

Light level meter.

Figure 4-8. Speed indicators and light meters.

 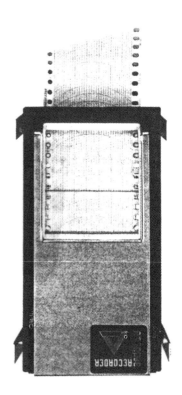

Clamp ammeter voltmeter.

Figure 4-9. Electric testers.

vibrating or surging machinery should be removed and tested at least every six months. Of course, digital equipment is available at higher cost, but don't forget to check the batteries (see Figure 4-7.)

• Speed—Instruments are direct contact or stroboscopic light flashing (see Figure 4-8).

• Lighting—Use light meter reading in fc (footcandles) (see Figure 4-8).

- Electric—Ammeters, voltmeters, wattmeters, and power factor meters are available (see Figure 4-9).

- Special Conditions—Air flow volume from a ceiling outlet cannot be measured without a hood or collecting box since part of the air is being sucked in the center. You can buy one which folds up when not used. When used, it must be held up against the ceiling and can be read directly. On high ceilings, this may require a ladder. Cost is approximately $2,500. I prefer to use a box of cardboard or fiberboard which was the original instrument when ceiling outlets were invented. I use two sizes, depending on how large the outlet—24"×24" or 12"×12." Although not quite as accurate as the folding unit, you can make this yourself. I mount my anemometer in the base opening, put two slide rings with set screws on the side, and hold it up with two 3' telescoping poles. I count the needle rotations and use a stopwatch to calculate air volume. Shape of box is in Figure 4-10.

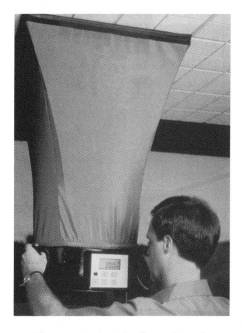

Figure 4-10. Air flow hoods.

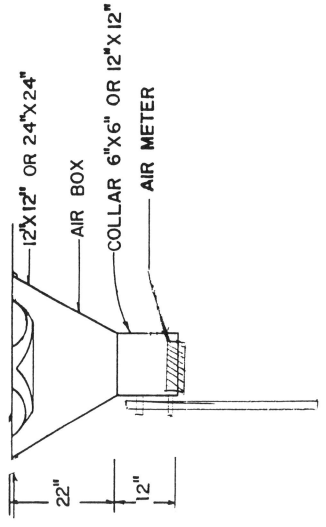

Figure 4-10 *(Cont'd)*.

CHAPTER 5

SICK BUILDING SYNDROME

CAUSES

Mold and Mildew

Mold and mildew grow on anything wet. How do building surfaces get wet? Dew point is the temperature when moisture drops out of the air. If the building is air conditioned, say at 75°, all of its surfaces are nearly 75°. If the outside air dew point is 80°, it is above 75° and quite often in the southern part of the United States, it gets up to 80°. Then, everything in the building can get wet if the air conditioning is not running.

Mold and mildew not only cause respiratory problems, but can damage and spot clothing, furniture, carpets, wall covering, ceiling panels, paper, air conditioning equipment, and ducts. They can even collect in the waste paper basket. In severe cases they can peel wall covering off and even penetrate walls and insulation.

Algae

Algae is also a bacterial material but grows in very wet areas like air conditioning drain pans, cooling towers, drain lines, and traps. Legionella has been found in cooling towers. If you want to find out if there is any algae in a drain pan, run your finger over it and if it is slippery, or feels like jelly, you've got algae.

Off Gases

Many materials like paint solvents, furniture, carpets, wall and ceiling materials give off objectionable gases. Formaldehyde and urethane used to be present, but recently have been outlawed. If your furniture or carpets are old, they may still be emitting these chemicals. You can test for these chemicals with a toxic gas detector (see Figure 5.1).

The Carpet Manufacturers Association recommends airing out a space for two days after new installations.

Office Machines

Copiers may emit ozone and carbon particles. Most computer printers are safe.

Temperature and Humidity

Most people are comfortable between 72° and 76° and 40% to 60% relative humidity. Since the average body temperature is

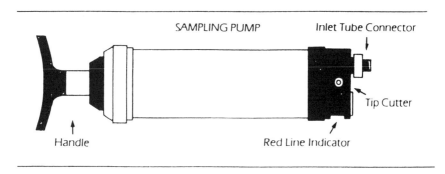

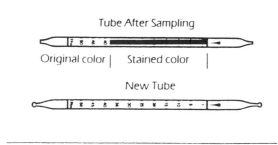

Figure 5-1. Toxic gas tester. Toxic gas detector tubes—complete and economical system for the detection and measurement of toxic gases and vapors.

98.6°, heat must be removed from body surfaces to stay comfortable. About 60% of heat is lost by radiation and 40% by evaporation, so the surrounding air must absorb this. Many persons' physical condition may require different temperatures. In Florida and gulf states, high humidity is a problem requiring lower temperatures. In dry western areas and in winter in the north, low humidity becomes a problem. Conditions encountered are allergies, respiratory distress, sweating, skin **irritations,** and general discomfort.

Air Circulation

Since body heat must be removed, the air surrounding a person must absorb this heat and moisture. If the air does not move, it becomes warm and saturated, so there must be some circulation.

Many new air conditioning systems use what is called variable air volume operation. When the area reaches the control temperature, the air circulation shuts off or is diminished. This makes some people feel like they are living in a vacuum. The change in air circulation also creates discomfort. Low air flow from some outlets lets the cold air drop to the floor on cooling and lay up at the ceiling on heating.

Cleaning Materials

Some materials used for janitorial purposes, such as chlorine solvents and waxes, may give off objectionable odors. These materials may be stored in air conditioned rooms, and can cause a serious hazard if they leak and the air picks this material up, circulating it throughout the occupied space.

Building Construction

The type of wall construction affects the feeling of comfort. If the surface is much warmer on the inside or colder than the room, your body will feel the radiation.

In a bank which had 16-foot-high thermopane solar windows in "double glass layers," the surface temperature on the south side at 4:00 PM was 135° with an outside temperature of 40°. The temperature on the desks five feet from the windows was 95°, the rest of the bank was 75°

Even mirrored glass gets warm when sun shines on it.

If the walls are masonry, you have about a 6-hour time lag from solar heat so that the surface is usually fairly stable and should not provide any discomfort.

Moisture through Walls

Even though the walls may be waterproofed, masonry, and are insulated, there is a possibility that if there is a negative pressure in the building, some air and moisture can be drawn through from the outside.

Most codes now call for a certain amount of insulation in walls but most insulation is made with a vapor seal on the *inside*. In high-humidity areas, that vapor seal should be on the *outside* to prevent moisture coming into the building. Many cases have been found were the moisture penetrated from the outside, turned the insulation completely black with mold, dampened the inside wallboard, and discolored and/or removed the wall covering.

About 40 years ago, a university in Indiana built a very well insulated house and did some testing with humidification up to 50% in the winter time. They had insulation in the ceiling above the second floor and the building was insulated in the walls. When spring came, the second floor ceiling collapsed since the moisture in the building had penetrated the drywall and frozen in the insulation above and the weight of the ice broke through. The moisture concentration was so great in the basement that there were branches growing out of the floor joists. If you are going to humidify your building, be careful how high you make it and what materials you may have that will be affected by this high humidity.

Air Conditioning Equipment and Ducts

Much of the mold, mildew, and algae grows in the air conditioning system in ducts simply because these surfaces may be colder than the dew point of the surrounding air. If the equipment runs continuously and is properly maintained and cleaned, there is no problem. If the equipment shuts down at night and the building is not sealed up, that means that there is no outside air damper closing on the air intake or in the exhaust openings through the roof, so that moisture floats into the building. Those cold surfaces in the air conditioner and the ducts will get wet, mold grows there, and when the air conditioner starts up in the morning, it blows these contaminates all over the space. The drain pan, for instance, if it doesn't drain properly, can grow a lot of algae. Bacteria grows in this algae and can get into the system. Some of the materials which are placed in the drain pans do not always solve this problem unless the pan is properly coated, is not rusty, does not have a rough surface, and the drain is in the bottom rather than the end (see Figure 5-2). We have tested heat pump units which had an end drain and there was 3/8 of an inch

Figure 5-2. Ten-year-old air conditioning unit drain pan.

of water in the drain pan, all loaded with algae. Filters do not always take out the dirt, and few of them take out bacteria, so the coils and even the fan should be cleaned at least once every year or two, depending on how dirty the space is. If possible, you should use a better filter, usually a pleated type with a lot of surface. The best filter, of course, is the electrostatic type.

Outside Air

You may have contaminates in the outside air which may be dust from some other source, or can even be materials in the moisture blowing out of the cooling tower which may be alongside the air intake. Algae, germs, and other materials grow in cooling towers unless they are properly treated. Air may be contaminated from garages and adjoining highways which have a lot of automobiles spewing out carbon monoxide and nitrogen oxides.

Odorizers

There are many materials on the market with pleasant fragrances and items even plugging into electric receptacles to give off pleasant odors. Some people are allergic to these fragrances, so if you are using this type of material and you have people complaining of respiratory conditions, this may be the problem. We had a condition where a couple had two large dogs who were allowed into their bedroom. The wife used a fragrance sprayed on the carpeting to air it out when the dogs were gone. During the night the couple seemed to have headaches and were perspiring. I ran tests for humidity and found no problem there. I called the manufacturer of the spray who said there was only boric acid, starch, and a fragrance in it, but that certain people were allergic to fragrance. I suggested eliminating the use of that material, and taking sachets out of drawers. Their problem went away.

Noise

Noise is a general irritant, especially if it is continuous and it's the wrong frequency, often caused by vibration, unbalanced fans, and equipment which is not properly installed. Also noise from outside sources should be reduced.

Lighting

Lighting has been the subject of discussion for years. The Illuminating Engineering Society has set up certain light levels for proper use and these should be complied with. Today with so many computers **in use,** there are many problems with headaches, glare, and fatigue from using this type of equipment. You should investigate where the lighting is, how it is arranged, and what you can do to improve it.

Stress

There isn't too much you can do about stress, since this is a function of the operation of a particular facility. The condition of an employee, problems at home, and problems in the work place sometimes cause sickness which is blamed on the physical environment. So you might investigate a person's condition thoroughly if stress is evident.

CHAPTER 6

TYPES OF CONTAMINANTS

In addition to algae, mold and mildew, other materials which will cause difficulty are methyl alcohol from spirit duplicators, methacrylate and ozone from copiers, sulphur dioxide from heating systems, pest control agents such as chlordane, and organic compounds such as carbon dioxide, carbon monoxide, sulphur dioxide, nitrogen oxides, ammonia, and radon.

Fibers in the air can come from clothing and other textiles, cotton, fiberglass insulation, and asbestos. Dusts comprise common lint, household allergens, house mites, pollens, dander, smoke from tobacco, wood, or coal. In stagnant water, as we mentioned in drain pans, and also in pools, plants, waterfalls and fountains you can get microbes, bacteria, protozoa, fungi, and viruses.

Even though plants are supposed to absorb carbon dioxide and give off oxygen which should improve the condition in the building, the moisture in the ground under that plant, may be generating pollutants by itself.

The World Health Organization estimates that 30% of all new and remodeled buildings suffer from "Sick Building Syndrome." EPA studies indicate that air inside buildings is often as much as 100 times more polluted than the air outside. With people spending nearly 90% of their time inside, it appears that sooner or later employees will begin to report difficulty with their health.

Heat Recovery Equipment

Some heat recovery equipment consists of a heat wheel in which exhaust air passes through the wheel, is cooled partly, and rotates over into the outside air intake which precools it before it goes into the air conditioning system. Quite often the exhaust system includes rest rooms and contaminated areas like isolation rooms in hospitals. If this is the case, there is a possibility that the heat wheel can become contaminated, carry contaminants into the outside air intake, and put them back into the air conditioning system. A special filter known as a "Hepa" filter which can screen out most of these particles. Hepa filters are not popularly used except in extreme cases because of their high cost and high pressure drop. A 2'-square panel may cost as much as $250.

A better system is using a metal coated system in which the exhaust air passes through one series of openings and is isolated by metal panels from the outside air entering. The heat transfer and the recovery is not as good, but it is a lot safer from a contamination standpoint.

Another location where germs and contaminants may be generated is toilet rooms if they are not thoroughly washed and disinfected often. Body fluids sometimes leak onto the floor and lay there until dry. In the meantime, they are giving off perhaps germs and contamination into the air in the room. If the room is not continuously exhausted, there is a possibility of collecting these materials in the air as well as on the surfaces in the room. From the standpoint of energy used, these fans should be shut off when the room is not used because the exhaust pulls outside air into the building, causing difficulties elsewhere. Therefore, it is up to your consideration to determine how to operate these fans, depending on the procedure for cleaning.

Quite often, elevator pits are not provided with sump pumps and they can store water and cause difficulty, especially on hydraulic systems which are used in fairly low-rise buildings. If you have this condition, it is recommended that you install a sump pump in that area to keep that space dry so those contaminants do not float up into the elevator cab and bother occupants during use.

CHAPTER 7

CURES FOR SICK BUILDING SYNDROME

CONTROLS

As mentioned previously, the control of atmosphere in various areas may help the feeling of comfort as you walk through a building. Energy management systems are very important today, but they should be so arranged that the maintenance man or even the building manager can look at the computer screen and tell if there is anything wrong. If he gets a call of complaint, he can get the answer quickly. If it needs maintenance, a maintenance person can then determine what is wrong, right from the computer without going to visit the site. The most important thing with controls is that there should be a training program and an operating manual written in the kind of language that a maintenance man and the manager can understand.

Prevention of Mold and Mildew Growth

The major problem causing this condition is having the building open during periods when the outside dew point may be higher than the inside temperature. This not only occurs in the south, but I have had it occur in areas near the Great Lakes. It seems that many of the systems **designed** have what is called an enthalpy or economizer control which brings outside air into the building to cool it off when it is colder outside than inside. This is fine as long as the outside humidity is not high and moisture does not enter the building, condense on cold surfaces, and grow mold which spreads throughout the space. Mold, by the way, looks like ripe dandelions with a bunch of spores which spread all over the place. We encountered moisture penetration a number of years ago in a system I designed for a furniture store in Ohio. I used the economizer system, but I didn't realize that at night, if it rained, it brought moisture into the building. One day I got a call from the owner telling me that all his furniture was wet. I quickly shut the control off.

Air Conditioning System Operation

As mentioned, if you shut the system down at night, you must seal up the building. If you have cleaning compounds used during the night by your Janitorial service, start your air system up in the morning without the cooling on if you can. It takes a special control. Get some air circulation and some outside air in if it isn't too damp outside. Exhaust the air if possible; that will get rid of the compounds, especially the odor from chlorine materials.

Air Conditioning Equipment Maintenance

You should clean the system out at least once a year. If your coils are cleaned by removing the coils from the air handling unit, washing them out, and replacing them, make sure that whoever reconnects the refrigerant circuit has evacuated the piping and reinstalled the refrigerant. Otherwise, you will get air bubbles in the system and the operation will be very inefficient. We have tested one 10-ton system where this had happened and the equipment was running at 60% efficiency. Don't forget to clean the fan, because if those little curve blades get plugged up with dirt, they can reduce the air quantity down to almost 10% with the result that the cold air coming out of the air outlet may be so low that mold and mildew will grow in the duct and on the air outlets.

As mentioned, gases are given off by certain types of machinery such as copiers. Place an exhaust grille near the source of these gas emissions and exhaust them into the return air system. If, like Ozone, it is a more dangerous gas, exhaust it to the outside. If these gases come from furniture or other materials, you may have to bring in more outside air and replace this air until this concentration is reduced, or you may need to replace the materials giving off these gases.

Air circulation is critical to a person's comfort. Air velocity, of course, becomes a problem when people are exposed to drafts. Use of ceiling fans helps air circulation, and occupants can tolerate velocities as high as 300 feet per minute, it is possible to raise the temperature as much as 6° and still feel the same level of comfort.

WATER TREATMENT

Water treatment systems are extremely important because nearly all water in the country has some minerals in it. We tested a 4-year-old office building which had a water treatment system in the cooling tower circuit, but it had not been properly maintained. The result was that the cooling tower louvers had practically all rusted out, and the cooling tower was half plugged up. The piping system in this 16-story building was partially plugged up with lime. Since the building was not entirely occupied and the heat pump units had not been installed in about 5 floors, sections of the piping were removed for inspection. It was found that on at least 3 floors the piping was entirely plugged up and there were pin holes in the piping which were leaking.

On another project consisting of a group of apartments in a number of buildings with underground piping, the shower controls were not working because they were plugged up with lime. Examination of the system indicated that the city water was coming in, going through a hot water boiler—good sized one—then circulating to the various buildings underground without insulation. This particular problem was solved by removing the control unit from each shower valve, replacing it, and then taking the original unit and cleaning the lime by dipping it in a lime-removing solution. Also, to prevent the boiler from liming up, a heat exchanger was installed using circulating water from the boiler into the heat exchanger and a separate circulating pump to pump the domestic water through the heat exchanger to the various apartments.

Improper water treatment in outside cooling towers or evaporative condensers may also support the growth of dangerous bacteria in the drain pan, which may be sprayed into the air, getting into the outside air intake.

LIGHTING

With lighting arrangements it is quite important to provide the proper amount of light at workstations, and in a direction that will not cause interference with vision. Work at computer screens or television monitors requires special attention to lighting. The types and color of lighting are important. Certain types of lights are objectionable. Incandescent bulbs create heat which may radiate down to an occupant if the bulb is close enough, causing discomfort. Certain high-density bulbs which give off good light may interfere with the surface appearance of some papers.

BUILDING REPAIRS

If you are in a building and part of it is being remodeled and repaired, it is now required that you seal off that section so the dust and dirt does not get into the occupied space. This may be quite difficult if a single air conditioning system is supplying both areas. This condition can also result in some legal problems if a person is affected by this dust and contamination.

Physical comfort is also discussed under the "Solution" areas, depending on the type of seating, standing, and the position of hands and feet during operations so that comfort is provided and fatigue is reduced over a period of time.

WORKPLACE ARRANGEMENT

If you are working in a confined space isolated from other people in the general area, you may feel confined, although you know that the interference with sound and voice from another area may bother you in your operations. Use of sound-absorbing materials and a limit to height of partitions so that the overhead area appears to be of reasonably good space, will help a person feel better in such an atmosphere. Here, again, air circulation is extremely important as persons who are in a confined area sometimes feel depressed if there is no air motion around them. This once was a major problem for people who smoked while they worked and the stagnant air remained smoke-filled. Today, of course, with restrictions on smoking, this may not be a problem.

CHAPTER 8

SOME QUESTIONS ON RESIDENTIAL HEALTH PROBLEMS

1. What happens when urethane plastics burn?
   Answer: Generates hydrochloric acid and phosgene.

2. Does algae grow in the drain pan of air conditioning units?
   Answer: Yes—use chlorine solution and flush out before running fan.

3. Do filters in residence A/C units get mildew on them?
   Answer: Yes—it looks black. Filters should be washed or replaced.

4. Does foam insulation in house walls and ceilings cause chemical emissions?
   Answer: Yes—if not properly installed could give off formaldehyde.

5. How about particle board?
   Answer: When this board is cut, formaldehyde is given off. Use very good ventilation if you are working with this type of board. New materials may not have this.

6. Are insecticides dangerous?
   Answer: Maybe, check the container. If you have exterminating service, ask them what they are using and what problems may occur such as contamination of exposed food. Chlordane is no longer used.

7. Are gas, propane, and kerosene heaters dangerous?
   Answer: They can be. They give off carbon dioxide, nitrogen oxides, and hydrocarbons. If not properly adjusted, they may

give off carbon monoxide which is deadly. Vent the fumes to the outside.

8. Are charcoal barbecue grills safe?
   Answer: Charcoal can give off carbon monoxide. Keep your head away from above the grill. If you use charcoal indoors, get good ventilation.

9. How much outside air do we need for ventilation?
   Answer: 5 cfm if no smoking, per person
             30 cfm if smoking, per person
             Bathroom fan exhausts 50 cfm
             Kitchen exhaust fan 100 to 150 cfm

10. Should we be careful in using lime and mildew cleaners?
    Answer: Definitely—read the box or bottle label and believe it.

11. How should I eliminate drafts from air conditioning?
    Answer: Turn grill blades up in summer, down in winter. Reverse ceiling fans.

12. What is the comfort zone?
    Answer: Between 72° and 76° and 40% to 60% relative humidity. In South Florida, to save energy, set at 76° or with ceiling fans, set at 80°.

13. Is keeping house or apartment wide open on mild, wet days a good idea?
    Answer: No—Moisture increase in the house may soak up very slightly in furniture and carpets. Although little is absorbed at one time, it may be cumulative.

14. If you have a basement, check for moisture on walls and floor. Test for radon.

15. If the air conditioner and open-flame water heater is in the basement, leaking Freon refrigerant can cause phosgene or mustard gas which is deadly.

CHAPTER 9

COSTS AND SOLUTIONS

1. Poor worker production. If a worker is uncomfortable, unhappy, or under stress, their production is lowered and this can cost quite a bit of money. The average costs of workers' salaries is $130 per square foot per year.

2. Absenteeism caused by poor health or conditions aggravated by the interior air circulation may cause workers to be absent an unacceptable amount of time. We had one case where an office had eight people, three of whom were home at least three days a week and one who worked four days a week, but it was necessary for her to wear a face mask in order to function.

3. Another source of problems is high energy bills. See the recommendations under the items that should be performed originally according to building specifications. Now look at items that were not included. Items such as shutting down air conditioning systems at night and on weekends, covering windows when the sun shines, and eliminating the cooling system during janitorial services can reduce your energy costs.

4. Damage to equipment and furnishings—some materials may damage furniture, discolor it, or produce odors which make it uncomfortable to use. There have been cases where high humidities have sealed envelopes, damaged paper, and copying machines, and even peeled wall covering off of walls. If you go into a building with a lay-in ceiling, look to see if the

panels are flat or are sagging. If they are sagging, that building has had high humidity and those panels were affected by it. This may require replacement of the ceiling panels to make the building appear as originally designed.

5.   Poor interior conditions such as high or low temperatures cause discomfort or irritation. In retail areas, customers may not stay long, and sales may be lost.

6.   Initial inspection and tests—do these first as described under Chapter 4 to see if you can find out what the actual problems are before calling in an expert. You may also determine what may be causing the difficulties, as well as methods of correction, from the description of the building operation.

7.   Computer management systems—there are now instruments available on the market to sense certain contaminates in the air like carbon monoxide, carbon dioxide, and even dust particles. They can be measured and determined before the manager or the maintenance man responds to a complaint.

8.   Remote control—a computer or other device may also provide a remote control that can change the conditions of the air in a space so that personnel do not have to visit the area or come in contact with the occupant who has difficulty. If the computer is properly designed and there are sufficient sensing devices in a building, conditions can be determined by good investigation and a proper training manual.

9.   Alarm sensors—in that same system, alarms can be provided, for instance, at a security office, in the manager's office, or at his secretary's desk, where a flashing light could indicate a problem. The alerted person could check the computer for more information, or call a maintenance man, to find out what is wrong.

10. Prepare an operation manual and program for your specific building, using the original information from the engineer and architect and the control contractor. If such is not available, it is necessary to inspect and test the entire system to find out how it is operating, what is necessary to make it work right, and if you wish, how to design a management and detection system to take care of undesirable conditions when they occur.

CHAPTER 10

SPECIFIC OCCUPANCIES

OFFICES

These areas are difficult to control and satisfy all personnel because most of the work is sedentary and the characteristics of each person's metabolism change from day to day. It is necessary for the person in charge of a particular department to set down rules and regulations as to what the temperature is going to be, attempting to average it at a reasonable point for everyone.

OUTSIDE CONDITIONS

Cold, rainy days may require higher heat settings in the morning than in the afternoon. Very hot days may require slightly cooler indoor temperatures until personnel become adapted to the work environment.

If a building is large, there should be separate heating and cooling thermostats for the interior, since this center area only changes due to internal load, not in accordance with outside exposure. If there is an area which is used intermittently by large groups of people, such as an auditorium or meeting room, or if there are areas which have heat-producing machinery, these should be arranged so that the air supply can be changed in volume or temperature or can be shut off when not in use.

In winter, furniture will cool off and it may, take as much as four hours to reach ambient temperature if the temperature falls too low. Your personnel should be advised that this is necessary for energy reduction.

If persons are seated near outside walls, solar energy in summer, cold wall radiation in winter, and leakage due to wind greatly affect comfort conditions. Unless provisions are made by the air handling system, by heating devices or cooling devices along the wall, personnel sitting within three feet of an outside wall may be seriously affected by these conditions. Reducing the cross-sectional area of the body exposed to an outside wall such as moving from back exposure to side exposure may help reduce the uncomfortable effect of an outside wall. This discomfort will again require lower temperatures in summer and higher temperatures in winter.

Arrangement of personnel, office fixtures, sound baffling, and partitions should be examined for their effect on the air flow through the space. The air pattern from certain types of ceiling outlets and wall grilles can be seriously affected by partial partitions. These confined spaces may require lower temperatures in summer and higher temperatures in winter.

AUDITORIUMS

In auditoriums and assembly halls, due to the rapid change in cooling and heating requirements, oversized equipment is usually necessary. It is suggested that the air be pre-cooled or pre-heated for at least 1/2 hour before occupancy and be immediately shut down after use is completed. It is also suggested that two-speed equipment be used or other means of capacity reduction, permitting the system to operate at minimum capacity during low-occupancy and at full capacity during occupancy.

If considerable stage lighting is used in the auditorium, such as that required for drama or large stage performances, examine the possibility of re-use of the heat from this equipment into the heating system of the auditorium either for make-up air or for general space heating. If a large quantity of lights are used near the ceiling, test the ceiling temperature to determine if the air at the ceiling can be recirculated to the floor in the winter or should be exhausted to the outside in the summer.

GYMNASIUMS

Ventilation
Play Area
The general play area and seating area of a gymnasium should be provided with a remote outside air control so the quantity of ventilation and the quantity of exhaust may be varied in accordance with requirements of the number of people in the space. This is particularly valuable during periods of high occupancy such as during basketball games where more ventilation is required

Outside air can be used for cooling during large occupancy if outside air is below 60°.

Locker Rooms
Since locker rooms are normally required by code to have a high rate of ventilation because of odors and contamination, means should be provided for supplying this area only during periods of occupancy. During other periods requiring either heating or cooling, the air should be recirculated in these spaces without outside air being supplied.

Drying Rooms
Areas in which uniforms are dried should be provided with heating devices for wintertime drying. For summertime drying they should be supplied with an outside air supply and exhaust for using outside air. Drying rooms should be enclosed and reasonably tight fitting so that the heat does not dissipate into the adjoining locker room or space, reducing efficiency and using more energy.

Physical Therapy Equipment
Whirlpool tubs and other equipment requiring hot water or heating devices should be carefully controlled as to their operating periods and they should be shut off when not in use.

SWIMMING POOLS

Water Heating

If pool water is heated, the heating device should be a storage type or a continuous heating device having modulating control. Periodic on and off control of large-size heaters for this type of operation is inefficient. Water heaters should be carefully examined for condition of the heating surfaces. Look for minerals on the water side and soot or dirt on the heating side, if fuel fired.

Water Filtering

Filtering systems should be such that the pressure drop through the device is fairly low so that the electricity for pumping can be as low as possible.

Ventilation

Outside air is required by code to be supplied to indoor pool areas and should be admitted only in that quantity necessary. Construction of the pool room determines also the temperature and humidity of this space.

Humidity Control

It is not necessary to run the ventilating system at all times. A humidity control should be provided to shut off the ventilation system when the humidity is down to certain levels. The level should be determined by the construction of the building to eliminate deterioration of the building or to eliminate condensation of moisture on floor surfaces. In cold weather, this is very important as it will reduce the requirements for heating outside air to be supplied to these spaces. (Hospitals, hotels, and apartments are covered in reference books.)

REFRIGERATION

Freezers

Freezers should be well insulated and if they are used often during the day, should have two sections, an active section in the front and a storage section in the back, with intermediate doors. Examine the door gaskets, especially on the bottom, to determine if there is any leakage. Examine the refrigeration unit for dirt or ice coating which reduce efficiency of the cooling unit and waste energy by requiring the compressor to run longer. Feel the outside wall. If it is cold, insulation may be bad or leaking.

Coolers

Coolers should be examined in the same manner as freezers. Interior lighting of freezers and coolers should be provided with a pilot light and switch on the outside so that lights may be shut off when not in use.

Combination

If a combination cooler/freezer is used, the freezing compartment should be well insulated with a tight-fitting door, to prevent moisture penetration into the freezer. Moisture on the refrigerating unit causes frost or ice buildup and reduces efficiency.

Refrigeration Equipment

Refrigeration equipment should be located preferably in a remote area or separate room so that if air-cooled, it can be provided with outside air in the summertime, and the heat from the refrigeration can be re-used in the heating system in the wintertime. If the refrigeration is self-contained within the cooler or freezer, it is normally located on the top or the bottom. The condenser coils should be carefully examined for cleanliness. Dirty coils can add as much as 30% to the operating energy use of refrigeration equipment. (Provisions for kitchen, laundry, and similar facilities are covered also in one of the reference books.)

Domestic Hot Water Temperature

Normal operating temperature for domestic water is 140°. The water temperature at the faucet should not be over 110°. If the water piping is not too long and the losses are not too great, it is possible to reduce the water heater temperature to 110° to 115°. This reduces radiation losses from the heater storage tank and piping.

Domestic Hot Water Circulating System

If the piping system is lengthy, especially if it goes several floors, it is highly recommended that a small pump be installed on the system to keep the water circulating through the piping at all times. If the piping is long, it causes considerable loss of heat before the hot water arrives at the plumbing fixture and will waste not only pumping energy but hot water because a lot of hot water is wasted before it gets up to the required temperature. All hot water piping should have insulation 1/2" to 3/4" thick.

Lavatory Faucets

Spring-loaded or self-closing faucets with time delay should be used to prevent leaving the faucets open at all times. From the standpoint of energy, only the hot water really needs to be shut off, but if a self-closing system is going to be installed, both hot and cold should be considered. If this is not practical, or there is objection from personnel, at least some means of shutting off water flow in the main during periods of non-occupancy of the lavatory would help.

CHAPTER 11

EXAMPLES OF
ESTIMATED RELATIVE COST SAVINGS

These estimates are based on general information and experience and an electric cost of $.07 per kWh and natural gas at $6 per
MCF. These are not exact figures, but are presented to give a general range of cost possibilities for this type of operation.

I. COMBUSTION

   1. Combustion efficiency of fuel-fired heaters and boilers.
       10% increase in efficiency saves $1.04 per 1,000 pounds
       of steam generated

   2. Boiler condensate return.
       Save $6.60 per 1,000 gallons returned.

II. AIR CONDITIONING AND VENTILATION

   1. Cooling air supply temperature.
       Save $266 per year per 1,000 cfm supply of outside air
       by raising inside temperature from 72° to 78°.

   2. Air condenser condition.
       Dirty coils cost $4.66 per year per ton.

   3. Oversized compressor and hot gas bypass.
       Cost $21 per year per ton.

4. Heating operation period.
    Set back to 55° saves $.42 per square foot of floor area per year.

5. Central computer monitoring system on air conditioning and heating.
    Save $650 per 1,000 square feet of building per year.

6. Clean filters.
    Save 1.2 million Btu per year per kW fan rating.

III. DOMESTIC HOT WATER AND TOILETS

1. Hot water faucet
    Open hot water faucet loses $.75 per hour. Leaky hot water faucet loses $1,476 per year.

2. Toilet exhaust.
    Save $180 per year per 1,000 cfm by shutting off exhaust on nights and weekends.

3. Shower flow control.
    Reducing water flow from 10 GPM to 3 GPM saves $1.18 per hour per shower head.

IV. SHIPPING AND RECEIVING DEPARTMENTS

1. Shipping doors.
    Open shipping door costs $675 per square foot of door per year based on 4 hours per day opening.
    a. Door seals.
        Truck closure saves $2,100 per year per door based on 4 hours per day use.

V. INSULATION

1. Roof insulation.

Save $180 per year per 1,000 square feet of roof with 2"
added insulation.

2. Wall Insulation.
   Insulation on walls saves $180 per year per 1,000 square
   feet of wall with 2" added insulation.

3. Pipe insulation.
   Average 2" pipe, saving 480,000 Btu per year per lineal
   foot.

4. Exposed air conditioning duct insulation.
   Save 73,000 Btu per year per square foot duct surface for
   1" thick insulation.

VI. LIGHTING

1. Average light level.
   Save $261 per year per 1,000 square feet of floor area for
   each watt per square foot reduction.

2. Electric signs
   Save $261 per year for each 1,000 watt reduction.

3. Remove two 4' fluorescent tubes in a 4-tube fixture.
   Saves $56 per year per fixture.

4. Sectional light switching.
   Shutting off half the lighting during operation in 10,000
   square feet of warehouse saves $3,630 per year.

5. Transformer heat.
   Save $4,473 per year by recovering waste heat from a
   1,000 kVA air-cooled transformer.

VII. BOILER BLOWDOWN

1. Use of blowdown heat exchanger.
   Save 1,330 Btu per year per 1 million Btu per hour output.

2. Reduction in blowdown period.
   Save 266 Btu per year per 1 million Btu per hour output for a saving of 1 minute blowdown time.

VIII. BOILER OPERATION

1. Cleaning of boiler tubes.
   Save 1,330 Btu per year per 1 million Btu per hour output.

IX. WINDOWS AND DOORS

1. Windows.
   1" insulation over windows saves $333 per 1,000 sq. ft. of glass.

2. Replace broken panes.
   Save $50 per year for each 1 sq. ft. of opening.

3. Skylights.
   1" insulation saves $360 per year per 1,000 sq. ft. of skylight glass.

4. Gravity roof vents without dampers.
   Loss $25 per sq. ft. of opening per year.

5. Power roof vents operation shut down.
   Save $250 per 1,000 cfm fan capacity per year.

CHAPTER 12

NEW BUILDING DESIGN CORRECTIONS

These suggestions for additions or corrections of drawings and specifications for HVAC systems is based on 60 years experience by a consulting engineer. Most architects and engineers and even building owners sometimes cannot anticipate the operations of a building, especially if it is a multiple-tenant type like an office building. These changes and additions can help reduce operating and maintenance costs, and improve the comfort conditions of the occupants or process. This consulting engineer has designed the mechanical and electrical facilities for $1 billion worth of buildings and has field tested over 200 buildings and processes, resulting in estimated annual savings of $95 million a year. Most of the additions or changes in the specifications and drawings can be provided at minimal or no cost. As an example, a 50,000-square-foot office building is saving $5,000 per month for an investment of $5,000. Another office building is saving $200,000 per year for a minimal investment of $30. An air pollution problem causing several workers to be sick was solved by adjusting air outlets to the proper air supply volume. Oversizing air outlets for the amount of air being supplied had caused the problem.

With some engineers, the specifications—for instance on control—are very general, and usually do not explicitly state what is intended in the way of temperature, humidity, air flow, pressure, and controls. This problem is sometimes left to the contractor, who in turn leave it to the control contractor. He has to guess what the engineer had in mind, and usually ends up with a poor operation with very little or no proper instructions. The "user's" manual that the contractor supplies to the owner is a series of cut sheets from the manufacturer of the individual parts, with no general instructions on how the system should work, or what the operating conditions should be. It is typically not written in layman's language. How would you like to be a maintenance man who has to walk across a soft tar roof in summer to get to the roof-top air conditioners and come back with a layer of tar on your shoes? What about the service engineer who can't get to a "V" belt drive on an air handler because the engineer did not leave enough room to get to the opposite side of the unit from the entrance?

There is a lot of good equipment on the market. Improperly installed or applied, it can cost the owner and operator much grief. We have had a number of cases where air-cooled condensers have recirculated discharged air into the intake, resulting in 65% efficiency. We have seen boilers operating at 8% efficiency for lack of proper combustion. In one case a hospital with $250,000 worth of computer control equipment which was not providing control in most of the building. I have been in dining rooms where the thermostat was changed by six different people in one hour; in a large theater where the temperature was controlled by flipping a switch on and off in compliance with complaints from occupants; and where lack of air circulation created a health problem due to concentration of ozone in a copying machine.

Chapter 13

Maintenance Access

A lot of equipment is installed in spaces which are too small to allow proper maintenance. This is particularly true where the architect squeezes the engineer down to the smallest space possible so that the maintenance man cannot get, for instance, to the opposite side of a unit to change the belts or adjust the drive. Sometimes you can't even get in to replace the filters. In hotel rooms, and sometimes in similar small rooms, the unit is mounted above the main entrance. Most of the time, the access door to get to these units is either too small to access the entire unit or is not located under the unit so that equipment cannot be properly maintained.

To give an example of how important this is, one hotel with 44 rooms decided to replace all of the units after 5 years because they did not cool. Field tests and examination showed that there was nothing wrong with the units except that the fans and coils were plugged with lint accumulated over time from bed clothes and interior materials. With proper access doors, especially to the fan, they would have saved, in this particular case, at least $25,000 and perhaps more. Heat pump units are rather popular today for office buildings where there is multiple tenancy. The same is true if proper access is available to this equipment; it will be maintained properly, kept clean, and have reduced operating costs.

Showing the importance of access panels big enough to get into the fans and coils is the case of a 32-year-old apartment house right on the ocean in Florida. Normally, the coils with aluminum fins are corroded out in this period of time; but in these particular units, the aluminum fins looked like the day they were made. I asked the maintenance man why they were so clean and he said,

"I wash them every year." These units were located in small rooms but they were accessible. The fact that the units did not have to be replaced saved approximately $800,000. If you have heat exchangers like chillers, water-cooled condensers, and that type of equipment, make sure there is room to pull the head off and clean the tubes since most water leaves deposits.

This is also true of boilers so that you can clean out the boiler tubes and, in the larger type, get access to hand holds and boiler drums.

Chapter 14

Equipment Location

Rooftop equipment such as cooling towers, air-cooled condensers, and air-handling units and duct work should be elevated above the roof at least 12" to 18" depending on the size, so that the roofing can be properly serviced. We had one case where the air conditioning units were sitting on blocks of wood placed directly on the roof in a strip shopping center. Fifty-six units were on the roof. Nearly every one of them leaked after 5 years. The units had to be disconnected, raised up, and new pads put under them. This is also true of rooftop duct work which should be on elevated supports attached to a plate pinned to the roof for prevention of penetration (see Figures 14-1 and 14-2). There should be some roofing access such as a walkway from the location of the roof access to the equipment so that the roof may be protected from damage, and the maintenance man's feet will be protected from the tar.

City codes may require an enclosure around equipment on the roof. This should be properly installed so that it does not affect the enclosed equipment. For instance a ground-level cooling tower was enclosed in a perforated concrete block wall with about 50% opening. The enclosure was so close to the cooling tower that the 1,400-ton tower could only produce 1,100 tons (see Figure 14-3). In another case an enclosure which had a completely open louver on one side was affected by wind blowing across the top forcing the discharge air from a 200-ton air-cooled chiller back into the intake raising the temperature 8° and reducing the efficiency down to 65%. If there is a possibility of this recirculation due to wind extend the discharge up about 5 feet with a collar or increase the air velocity to about 3,000 feet per minute which should

Figure 14-1. 10-year-old leaking roof at duct connections and unit supports.

Figure 14-2. Repaired roof with seals at roof surfaces.

project the discharge air up about 20 feet and reduce the recirculation (see Figure 14-4).

If the self-contained air-cooled unit is inside the building and the condenser intake and discharge goes through the same louver make sure that the velocity for the discharge is high enough to get the warm air away from the suction side. We tested units in a

Figure 14-3. Cooling tower enclosure 1,400-ton capacity reduced to 1,100 tons.

building which had vertical architectural louvers where the intake air was 107° on a 90° day. This again reduced the efficiency of these 10- and 20-ton units to down around 60%. Check with the architectural design and if those louvers are so arranged as to restrict air flow, see if you can get the architect to change the louver design or find some other means of providing condenser air.

You should specify in designing the air handler that there be complete access to the interior, especially the coils and the fan, particularly if it is a forward curve blade fan. We have bad cases where dirt in the coils and the fan reduced the capacity of a 7,000 cfm unit to 800 cfm resulting in the need to run 24 hours a day at full capacity in order to provide cooling for a one-hour dinnertime meal.

It seems to be a custom to use the equipment room as a collecting area for return and outside air (see Figure 14-5). We have

Figure 14-4. Cooling tower high velocity discharge collar.

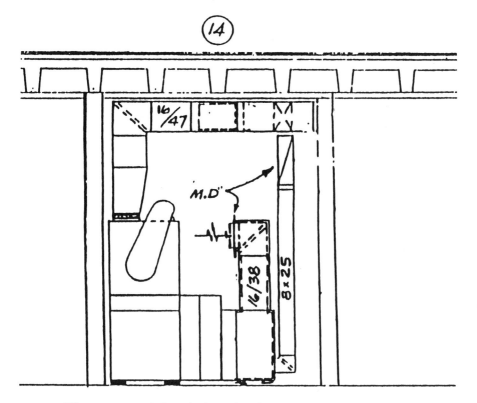

Figure 14-5. AC unit in mixed air room. Poor access.

had a number of cases where these rooms were used for storage of materials like gasoline, lawn mowers, and cleaning liquids and compounds. You should make direct connections with duct work to the return air and the outside air louver with proper dampers so that the room if it is used can be isolated from the air stream (see Figure 14-6). If not place a sign on the door prohibiting storage.

If the architect will not give you enough room for the equipment hang the unit outside the window. Show him the picture in Figure 14-7 and see if he can stand for this type of equipment on the outside.

This may not seem to be an important item but if there is any possibility of contamination in the air and the owner of the building is sued you could be in turn, sued for improper design of the system. I have been an expert witness m court in a number of cases and can almost assure you that you will lose in this type of a case if there is any injury to occupants.

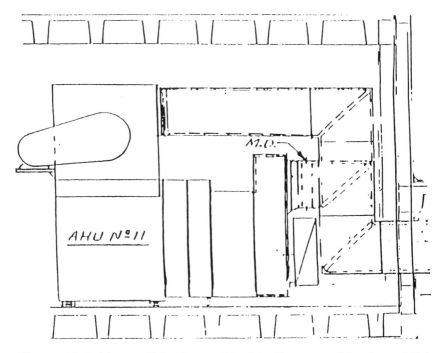

Figure 14-6. Air conditioning unit with direct outside air and RA connections.

Figure 14-7. AC unit outside building.

CHAPTER 15

System Design Recommendations

On air conditioning unit drain pans where it is required to put in a trap to prevent loss of air to the drain put in clean-out plugs on both sides of the trap to allow for cleaning. We have tested a number of units where the trap was completely filled with algae and the drain pan overflowed causing damage to ceilings and floors (see Figure 15-1).

With pumps, put the check valve on the discharge side since pumps are susceptible to low NPSH. There should be some positive pressure on the suction side of about 4 pounds per square inch (psi) for a 1750 rpm pump and about 10 pounds psi for a 3500 rpm pump. Also if the pump is on the condenser system of a cooling tower don't forget to add a strainer which is easily accessible and easily cleaned between the cooling tower sump and the pump (see Figure 15-2). We have seen dead birds inside a cooling tower where the intake was covered with feathers. We have also found sand and contamination due to poor water treatment in the tower which should have been picked up in the strainer. Put thermometers and pressure gauges in your lines. A gauge on the suction side of pumps should be a compound type which indicates either a vacuum or pressure. If you don't have a sensor to a computer these gauges should be provided with shut-off cocks and if left in the on position they should be replaced about every 6 months because the vibration usually damages the gauge.

To show the importance of a thermometer, for instance, the system for a 20-story, 300,000-square-foot building in the Midwest was having trouble cooling the building. It was a 2-pipe system using chilled water through a heat exchanger in the summertime from an absorption system, and hot water in the wintertime from

Figure 15-1. Unit drain trap, change elbows to plugged tees.

gas-fired boilers. A look at the thermometer on a supply line indicated 82° part of the time. The problem in the summer was that there was a leak in the control valve on the heat exchanger on the chilled water line and steam was leaking through heating the water up instead of supplying chilled water (see Figure 15-3).

In boiler rooms make sure you have sufficient combustion air intake. We had one building where four 10-million-Btu-per-hour gas boilers were burning the gravity gas burners out every year. We checked the combustion air and found it to have 15% carbon monoxide. We discovered there was no combustion air intake to the boiler room and there was also an exhaust fan. This was corrected by reversing the exhaust fan. Another case where a pressure gauge would have been helpful involved a gas boiler feeding an absorption system where in June the measurement of flue gases

Figure 15-2. Cooling tower discharge strainer.

indicated 10% carbon monoxide. Investigation indicated that the gas pressure on the burner was 15" instead of the normal 4." It seems that the previous winter gas pressure got low the operator screwed down the pressure-reducing valve and forgot to back it off in the summertime. Fortunately this was a rooftop room where he could open the door or he may have died.

It is a good idea to supply dining rooms and similar rooms

Figure 15-3. Chilled water system indicators.

which are just partially occupied during the day with dual units or two-speed units so that only the part that is occupied is cooled. We have reduced the operational cost of this type of equipment by 35% with daytime control just by supplying the area with two units rather than one (see Figure 15-4). This is also true of computer rooms which must be cooled night and day. For safety's sake install two similar size air handling units for these rooms with an automatic switch-over. We have also saved about 30% operating costs in a vocational school which was run in the summertime in scattered classrooms. This required two major air conditioning units to supply the areas. By cross-connecting the duct work from one zone to another we were able to shut down one of the units and operate on the other.

Be specific about the use of flexible and fiberglass duct work which of course has a higher static pressure than sheet metal. We

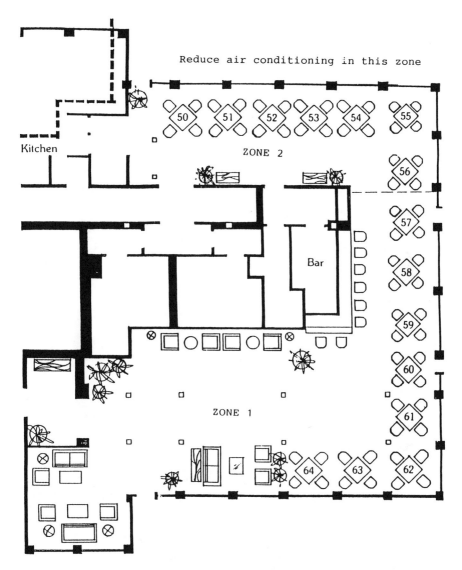

Figure 15-4. Dining room.

had one case in a residence where the contractor had to go from an 8" flexible to a 15" rigid fiberglass duct in order to supply sufficient air (see Figure 15-5). We have also had a residence where the contractor used flexible duct runs of 40 feet resulting in a discharge static pressure at the unit of 1" when the static pressure should have been .15" (see Figure 15-6). In this case the contractor had to spend $11,000 to replace the duct work in his house to make the air conditioning system work.

On flexible connections to vibrating equipment such as engines which seem to rock when they start up, make sure the pipe flexible connections are properly located so they bend rather than stretch, causing early failure (see Figure 15-8). In many cases, particularly in apartment houses, flexible connections to individual air conditioners are made with hose or so-called flexible metal connectors. In one case, a second floor apartment connector on an air conditioning unit in a 22-story apartment building caused $22,000 worth of damage before somebody found out where the leak was. In that particular case, they had to drain down the entire building to change the flexible connection.

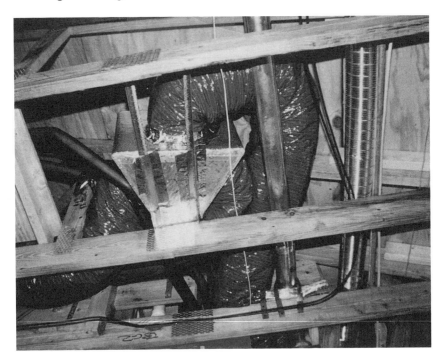

Figure 15-5. Flexible duct.

Figure 15-6. Fiberglass duct SMACCNA outlet.

On your pipe insulation, cover all cold and hot surfaces (see Figure 15-9). On your piping, make sure you specify a good seal coating on the outside, especially with rigid insulation like foam glass. We have had several cases where the seal on the outside opened up or there was no outside layer, the pipe rusted on the outside under the insulation, and the piping had to be replaced. Also, we have made measurements of similar insulation, not tightly sealed, on piping above a roof for chilled water. The temperature rise was 4° at 100' run which was nearly half the capacity of the system (see Figure 15-10). Also, it is a good idea to specify painting the piping a light color if it is for chilled water, and a dark color if it is for hot water as solar energy will affect the condition in the pipe. We have measured the surface temperature of black roofs at 90° outside with a surface temperature of 145°, and of white roofs at 95°.

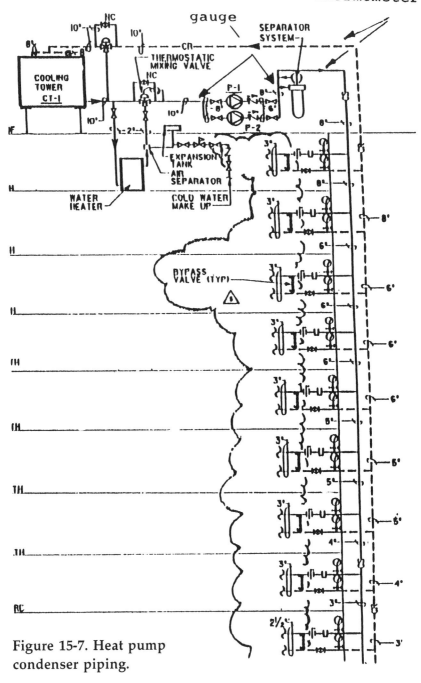

Figure 15-7. Heat pump
condenser piping.

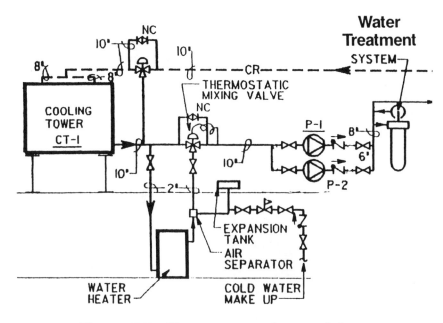

Figure 15-7a. Heat pump condenser piping.

Figure 15-8. Flexible connection for engine generator. Vertical ok, will bend; horizontal not good, will stretch and break.

Figure 15-9. Good pipe and equipment insulation.

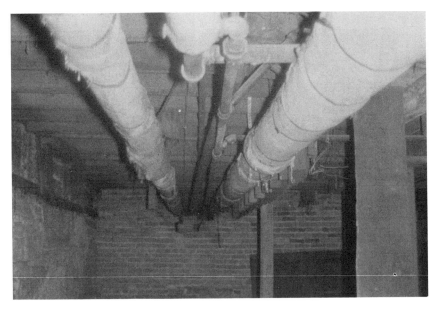

Figure 15-10. Failed pipe insulation.

We do not recommend the use of roof top exhaust fans for toilets and kitchens in high-rise buildings. We tested one 17-story building with this type of exhaust and found no exhaust below the 14th floor, even though there were dampers in the duct work. We also do not recommend the use of discharge from the fan into a common duct if individual fans are used, unless there is a back-draft damper properly operating. We have had a number of cases where exhaust from a bathroom on one floor ended up in a bathroom on another floor by gravity or by pressure, depending on the outside air temperature.

Also, when you specify an air handling unit, make sure the drain pan is large enough and the drain connection is in the bottom so that algae or any foreign materials do not collect and cause the drain pan to overflow. In nearly every case we have investigated, the drain pan has had algae growth in the bottom, especially in higher humidity areas. Even chemical pads, which are supposed to kill algae, quite often do not work.

If you are using a variable air volume system with a bypass into the ceiling space, do not use it under a roof since you are actually cooling the underside of the roof. Using a throttling type system, return air will be much warmer, and losses of cool air into the ceiling space will be reduced.

Another method of reducing operating costs is to provide a means of shutting off unoccupied areas in office buildings or hotels and apartments if fed from the same system We have found that in high-humidity areas, if the system is turned on just two hours a day in the warm or hot periods there is sufficient dehumidification to take care of any mildew or mold growth

If you are using a humidifier in a system put in a dew point minimum control to prevent possible injuries, especially in hospitals where you may be humidifying a surgery room. Water could condense in the duct and drop on the patient. If you are using an induction unit with no drain pan, water can overflow onto the floor because the supply dew point is too high. We have had a similar case where we almost lost two patients because of these conditions. Also use a minimum dew point control on an enthalpy system because, under certain conditions of high dew point outside, you may get damage to interior materials such as furniture or papers.

Put a water meter on the make-up line to the cooling tower or any other water-using system to make sure that there is not excessive overflow or leakage. Put one also on the make-up water to boilers since quite often, even with an automatic blow-down, there may be water loss in the system We have tested boiler plants with underground distribution piping where the loss was due to leaks in the piping and unrequired blow-down amounted to 20% of the fuel costs of the system.

On multiple-unit systems such as the individual heat pump units in office buildings, put a check valve in the condenser water line to the unit and preferably install shut-off valves on the inlet and discharge connection to the condenser. An example of the importance of this was an office building which had an excellent piping system with manual control valves and a flow indicator on each floor. It happened that two of the tenants in this building had computer rooms requiring one ton of cooling each. Since the building was all glass, it was not possible to cut a hole in the wall and put in a self-contained air conditioner. In order to get the heat pump units to operate, it was necessary to run a 20-horsepower circulating pump and a 15-horsepower cooling tower fan, nights and weekends, just to take care of these two tons. If check valves had been installed, the entire system could have been operated with two 1/3-horsepower circulating pumps located at the units, and the cooling tower fan could have been operated by a thermostat (see Figure 15-7).

If you are using a reheat system for control to get constant air supply, and you are using hot water in the reheat coils, consider the possibility of using cooling tower water at roughly 95°. We have made tests on reheat systems using 140° hot water. After running the tests to full heat, then to full cooling, we found that the 8" × 12" one-row coil took 20 minutes to cool down to the operating temperature when we switched the thermostat from heating to cooling.

If you are supplying air from a high ceiling, consider the possibility of using sidewall or dropping the air outlets to a lower level since it is not necessary to cool the air in the unoccupied upper space. We have been able to supply air from sidewall grilles in a church at the 11' level on an angle projecting up to 14' without disturbing the air from there up to the roof of 30', saving approximately 30% of the cooling requirements of the building.

In many spaces which are only partially occupied for certain periods, like conference rooms in an office building, why not use a separate fan unit or a fan VAV box so that the system can be supplied with minimum air and cooling during low occupancy and then switched over to high circulation during full occupancy? Air outlets should have some adjustment, be it by volume or direction, since the use of a space may vary from your original design intent.

Even though I have used air supply temperatures as low as 45° in comparatively dry areas, I have been restricted to a minimum of 55° in the high-humidity areas to prevent mold and mildew. The air pattern provided by the ceiling outlet should be carefully selected to make sure that the air gets down to the floor on heating and gets properly distributed at breathing level on cooling. Quite often we have measured a layer of cold air 3° colder than the breathing level near the floor, simply because the air was dumped down and did not distribute properly. This condition is quite prevalent in variable air volume systems where the air quantity may be considerably lower than the design pattern of the air outlet.

We had one case where several women in an office had respiratory problems because the air dumped straight down to the floor to a cold level. There was no circulation within this space and the air outlet was designed for some 150 cfm but was supplying only 60.

If you have a medium-pressure boiler plant, operating at approximately 100 pounds per square inch, investigate the possibility of going up to a 250-pounds-per-square-inch rating and using a steam turbine as a pressure-reducing valve, allowing the turbine to operate the feed water pumps or other equipment. You may find a considerable reduction in operating costs.

Do specify that duct connections to air outlets be made in accordance with SMACCNA recommendations, especially with fiberglass duct work. In testing 100 houses in the South, we found that 56 of them were leaking as much as 30% of their supply air where the duct connected to the ceiling drywall and the air was leaking out under the insulation. In one case where an attic space should have been 145°, it was actually 95°.

If you are using a direct gas fired make-up air unit, and you really don't need all that outside air, investigate the possibility of using a combustion chamber type unit even though it costs a little more. We had experience with this in a factory where switching from a direct-fired to a combustion air unit paid off in 1-1/2 years.

Although indirectly connected with HVAC, the amount of lighting does provide heat to a space and must be evaluated. If the wiring to lay-in fluorescent fixtures especially is made so that for four fixtures it can be done from a single junction box with flexible connections long enough to allow the fixtures to be moved, there is a considerable saving (see Figure 15-11). I did this in an office building which had a normal even pattern of lighting because the engineer didn't know where the people were going to be. I laid out individual desks for typing and computers, and located light fixtures directly over the occupants' areas, reducing electric consumption from 3 watts to 1-1/2 watts per square foot, with the same footcandle light level at the working space (see Figure 15-12).

A good example of the effect of lighting in an office building was a test that we made on two rooms exactly the same size, one above the other, with the same air conditioning unit type, and same supply air temperature. But one room had 200 footcandles of lighting and the other had 50 footcandles. The 50-footcandle room was 76° and the 200-footcandle room was 84°.

Figure 15-11. Typical lighting layout.

If the building has a lot of vending machines which are in a group, especially if they are refrigerated for cooling cold drinks or heated for other purposes, try to get an enclosure in back of the machines and return the heat generated to the return air in winter and exhaust it to the outside in summer.

Churches are good places to use thermal storage since they are only used part-time. Educational sections should be treated separately with a separate refrigeration and air-handling unit. Before getting into thermal storage, check with the power company as to their electric rates to see if it is feasible and if there are any real savings in demand charges by using this type of system.

Many buildings such as hotels, office buildings, and some apartment houses, have atriums. One problem with atriums is that normally you have to cool the entire area, especially if you have open walkways. The method of supplying air to these areas is

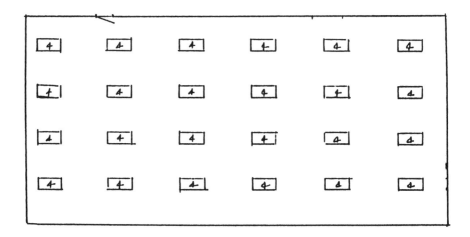

12 ft. Ceiling height—3 watts per sq. ft.

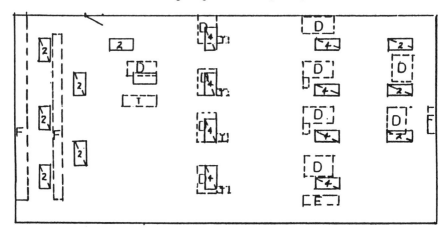

9 ft. Ceiling height—5 watts per sq. ft.

Figure 15-12. Lighting retrofit. 2- and 4-lamp fluorescent; d - desk; t - typewriter or computer.

very important. For example, I tested an 11-story hotel with a 300' × 100' atrium which had an air supply consisting of two large grilles, centrally located, above the 11th floor ceiling, blowing directly across into the atrium area. The temperature at the first floor lobby was 68° in the center and 91° and 92° at the other end.

There have been many discussions of how to handle atrium areas and there have been good systems designed; but, from the standpoint of energy savings and good air distribution, you might consider supplying the air at not higher than the second floor from both sides if possible. Use an air curtain type supply system down over the outside edge of the atrium to keep the open walkways cool by returning air from the ground floor up to the unit if it is on the roof, and by pulling air from the ground floor in summer and from the ceiling in winter.

Building lobbies are also important. If you have a large lobby occupied only by a receptionist, why not supply air with a separate control to the receptionist to keep that person in the range of 72° to 75° and keep the rest of the lobby about 78°? If you are coming in on a hot day, 78° feels pretty good!

We tested one building with an elevator lobby which was isolated from the rest of the building, and it was 68° when the outdoor temperature was 90°. The tenant rode the elevator up to his floor where the corridor was 70°, walked into his office which was 74° and said, "My God, it's hot in here!" He turned the thermostat down to 68° and after a little while he cooled down and turned it back up to 74°. He happened to have an office facing east and when the sun came out his back got war, so he turned it back down to 68°. Five minutes later his secretary came in wearing a sweater and said she was freezing. The tenant got on the phone and raised hell with the manager, saying, "There's something wrong with the air conditioning; you got to fix it."

One of the things that should be considered and should be told to an owner is that there is radiant heating off the windows regardless of the type of glass, solar or otherwise, even mirror glass, so that within 3' of that window you may get temperatures 3° warmer than the rest of the space. We have actually measured it in this condition. We have also measured solar glass on the west side of building at 2:00 p.m. with an outside temperature of 80° and a surface temperature of the glass on the inside at 135°.

You can provide controls separately in each of these places

and specify that they be operated in a staggered condition. In summer, set temperatures warmer in the lobby, a little cooler in the corridor, and let the tenant operate his thermostat. In winter, do the reverse; run the lobby about 65°, let the corridor be 70°, and of course the occupant sets his own temperature.

The other thing that is very important for office space is that the fans run all the time, for good air circulation. If you put fans on automatic, which is what you do in your house, you probably will get dead areas of air circulation, and complaints.

You can't always win. I tested one small office which had three women working in it and it was 76°. Two of the women were quite happy; the third had an electric heater under her desk.

Another thing you might look at, especially in high-humidity areas, is the operation of a fountain. If the water temperature is below the dew point, you probably won't have any difficulty. But if it is warmer, you may have evaporation which increases the humidity and causes your air conditioning equipment to run longer.

If the elevator equipment room requires cooling and it is in an area that gets warm, consider the possibility of using a small cooling unit recirculating air into the space, rather than an exhaust fan with an intake louver. We have measured as much as 4,000 cfm going up an elevator shaft because the exhaust fan was running and the dampers on the intake restricted the air flow. In this particular case, in a southern state, we used a 4-ton unit recirculating air, and closed up the exhaust fan and the intake in an elevator penthouse with three elevators. We saved about $1,800 per year with just this operation.

Most buildings today, because of the restricted outside air flow, have a negative pressure. This creates a number of problems such as air coming through the walls, through cracks in the building, and sometimes backing up on the exhaust systems. It is therefore recommended that the building be placed at a very slight positive pressure, especially in high-humidity climates where moisture may penetrate the walls, condense in the insulation, and create mold and mildew. Restoration requires an expensive program of tearing out the inside walls and replacing insulation. We recently solved an air pollution problem in an 8,000-sq.-ft. office space in a multiple-use building which had a direct opening to the outdoor air with no dampers on it. They shut the air conditioning off at night. By sealing the walls above the ceiling, and putting the office space on slight positive pressure, we were able to eliminate the problem. Five people, who had been sick and able to work only two to three days a week because of this problem, were no longer adversely affected.

It seems to be customary to supply outside air directly into the ceiling space of a building which is used for return air. If it is a high-humidity area especially, and the outside air is not cooled, condensation can occur, damaging the ceiling at the end of the duct feeding into the ceiling space, and creating contaminants. It is a good idea to supply the corridors and cool the toilet rooms indirectly by exhaust fans. This requires undercutting the doors to the toilets so that air can enter at the bottom and clear up the air in the toilet room. Figure 15-13 shows a typical layout where the outside air was supplied directly to the ceiling spaces, but could have been cooled by the units supplying the corridor and toilet rooms simply by connecting that supply to the ceiling space, omitting the air supply into the toilet rooms, and undercutting the doors as indicated. Another fallacy of this drawing is that there is insufficient duct size and capacity to take care of smoke removal and pressurizing the space in case of fire.

Also in the case of smoke removal systems, put back draft dampers on the supply fans and the exhaust fans so that air doesn't fall into the building when the system is shut down at night, creating overcooling and sometimes high humidities at night.

If you are concerned about freeze-up of heating or cooling coils, designers specify that the circulating system be such that there is proper water flow or refrigerant flow in the entire coil. We have measured large coils and found considerable variation in the

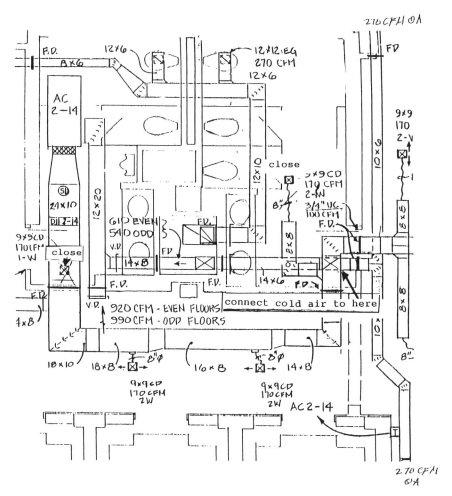

Figure 15-13. Uncooled outside air supply system corridors, toilets cooled.

air velocity throughout the face of the coil, in some cases it was slow enough to cause freeze-up.

If you are considering a cogeneration plant, check with the power company first to find out the cost and the arrangement of stand-by power to determine what equipment is necessary to ensure the power company's system is protection.

We encountered this in a proposal to provide generated power for half of a hospital since there were two separate electrical services to it. We checked with the power company who said that their equipment had to be protected in case of failure of the generators and the back feed. The protection was for a substation 4 miles away and would have cost $400,000 just for the protection for a generating plant which alone was only going to cost $600,000.

If you are using ground or well water heat pump systems, make sure an analysis is made of the water so you can specify the proper water treatment system if required. Most underground water systems collect a lot of minerals which create a coating on the heat exchange piping and, in some cases, put it out of business within 2 or 3 years.

It is a good idea to require labels on piping, designating what is in the pipes and what they are for. If the equipment in the building is to be painted, the piping should be painted a certain color. Most of the time the painting of that equipment is under the general painting specifications with no stipulation as to color, type of paint, or any labels on the piping and equipment.

Vibration isolators are essential, of course, on all moving machinery, especially in areas like penthouses where the space below is occupied, or the vibration may be transmitted into the building itself. We have found that mass accomplishes much more than just plain vibration isolators. In one 8-story building with all the pumps and the compressors on the floor above the executive offices, we used a 12"-thick foundation slab and then vibration isolators with good flexible connections on the piping, and there was no transmission of noise or vibration to the space below. Noise is very disturbing to occupants of many spaces which have ceiling-mounted units above the ceiling or within the space. Many times equipment must be turned down to its lowest operating speed. It is advisable that you specify certain sound levels of equipment to be used, since most manufacturers have this information. By specifying a low d.b. level, you may be able to offset future complaints from occupants.

On piping systems, be careful where you put the expansion joints, as, of course, all piping will move if it is at a different temperature than the surrounding area. This also applies to piping underground and buried in floor slabs or ceilings where movements of the building itself may cause difficulty. We have had the radiant heating piping snapped when the concrete cracked due to expansion, so that it had to be chopped out and expansion joints installed.

Specify good flexible connections on duct work, especially from the fan to the ducts if sheet metal is used, and also the method of connection to fiberglass duct work which is specified by SMACCNA and should be rigidly enforced. We have inspected a number of cases where fiberglass duct is just slipped over the housing of the fan discharge and has a strap around it—an effort to use the fiberglass duct itself as a vibration isolator. The duct work eventually breaks down and disintegrates.

Check with your structural engineer on particularly long, large buildings where your piping or duct work may cross expansion joints in the building, to determine whether that type of capability is required is case the building moves (see Figure 15-14).

Make sure to specify that any connection between copper and steel be provided with isolating or similar type materials so that there is no electrolysis generated. We have experienced considerable corrosion and loss of materials due to this. For instance, in a 100-unit apartment house, the domestic water turned red. Tests showed that a copper water line clamped onto the steel base of a hot water-to-hot water exchanger created sufficient current to remove about half of the tube sheet in the heat exchanger in one year's time. One milliamp at one millivolt will dissolve one pound of steel or iron in one year.

Specify on the operation of damper motors that they should reach the end of their stroke at 10% and at 90% to make sure that the dampers are tightly closed or fully opened at the end of the stroke. It is a very important stipulation to be made, as we have

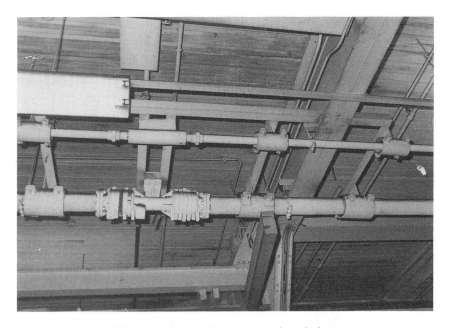

Figure 15-14. Pipe expansion joint.

had so many problems with leaky dampers, on outside air cooling and heating coil bypasses.

If you have multiple residential occupancy, and you are using the toilet exhaust to bring outside air in through the corridor from a corridor supply, provide an open space under the door so that the air can enter the apartment. This may cause difficulty with regard to smoke removal systems, so you must balance that advantage with the possibility of getting smoke under the door if the corridor is pressurized. If it is provided with negative pressure for smoke removal, than you have no problem.

When specifying air-handling units or any equipment with fans, make sure there is an access panel available to get into the fan so it can be cleaned, especially if it is a forward curve blade fan. If the fan itself cannot be cleaned from this access opening, a design should be made so that the fan can easily be removed for cleaning. If the unit has more than four rows of tubes and 18 fins per inch on the coils, it is a good idea to separate the coils sufficiently to be able to clean them. Close fin coils appear to be partially plugged up after about 4 or 5 years, even with good filter maintenance, and need to be cleaned. We have attempted to clean coils more than 4" thick with 100-pound water pressure and were unable to thoroughly clean these coils in a very dirty atmosphere. If you can, provide a space between the coils of at least 12" so that your water jet can get inside and clean them. Otherwise, make provisions with valves on refrigerant lines and steam and hot water lines so the coils can be removed and cleaned outside and then replaced. On refrigerant piping, valves are very important since the coil can be evacuated to remove any air and moisture, then reconnected to the refrigerant piping without disturbing the rest of the system. We experienced this on a 10-ton system which had the evaporator coil removed for cleaning and reinstalled. The refrigerant circuit had not been completely evacuated and recharged with the result that there was an air pocket in the refrigerant line resulting in a reduction of efficiency of about 60%.

Make sure you get good filtering operation, not the simple fiberglass mats, so that these coils are not plugged up. If there is an industrial or commercial process in the space, be very careful. A good example of oversight was at a plant which manufactured paper caps like they use in fast-foot restaurants. Unknown to me the sealer was paraffin. The air conditioning I specified for the project was using well water and there were 12 rows of tubes. The paraffin got into the coil and it was not possible to clean it; the coil lasted only 6 months. This can also happen in areas which have a lot of carpeting, heavily trafficked, causing lint to get into the system. We had a knitting plant which had this difficulty; the filter plugged up about every 6 months. We ended up using a continuous fiberglass filter. We let the lint collect on the filter, rolled it up, threw it away and got another roll.

For large industrial applications with duct work above the roof, you might consider the system we used in a million-square-foot plant where we installed more than 2,000 feet of 4'-diameter duct work by helicopter in one weekend. We used 20' lengths of ducts with a slip fitting and a flexible jacket. This particular duct handled 600° air for heating the building. The duct supports were installed ahead of time by cutting through the roof, welding the supports to the I-beams underneath, and resealing the roof ready to accept the duct work. The duct work was installed over weekends since the law did not allow helicopters to work above a plant if it was occupied by personnel (see Figure 15-15 and 15-16).

Figure 15-15. Helicopter duct assembly.

Figure 15-16. 600° duct connection.

If you do not have an energy management system and you have areas like laboratories with exhaust hoods, kitchens with exhausts, or other areas with a lot of exhaust, put a switch and a pilot light next to the door leaving the space so that someone can shut off the air conditioning or the exhaust, or even the lights from that one spot. Put the pilot light outside the door so that anyone walking by can make sure that space is closed. We have been in university science labs where all the hood exhausts and the 100% outside air cooling supply were going on in unoccupied rooms.

Be very careful with the air conditioning supply to libraries so that it is not too low, causing condensation in the books or, causing high nighttime humidity to damage books. We have had two cases of high humidity at night causing library books to come apart because there were no dampers in the outside air supply or the roof ventilators, at a university in the Midwest. Also in a library in Florida the supply temperature was 52°, which caused condensation in the books.

In museums, where temperatures and humidity are critical, is it really that important to have much outside air coming in? Provide just enough to have a slight positive pressure in the building to keep out excess humidity if necessary. This space is so large in comparison to the number of people in it that you really don't need that much outside air.

If you are designing a system for a bakery where bread is normally cooled by running it back and forth through a conveyor within the room and where most bread comes out of an oven at 400°, see if you can use the heat from the bread to provide warming of outside air supplied for ventilating and for combustion purposes. If you are using this system, consult the bakery to determine whether it is possible that yeast spores are not present to contaminate the air or the dough which is rising in the mixing area. An example of this system was a bakery making 16,000 loaves of bread an hour. There was a 60,000 cfm exhaust fan pulling air across the bread conveyor to cool it, dumping the air out through the roof. At the other end of the room was a 60,000 cfm direct-fired gas heater pumping heated warm air into the room. By using a recirculating system, the heating of the outside air was almost reduced to zero. In addition, in mild weather, there was sufficient heat generated in this operation to heat three additional floors in this plant.

Also, in laundries, control the rate of hot water fill into the washer, and with storage of hot water. Uncontrolled flow rates will cause short-cycling of the boiler and increase fuel consumption.

If you are designing athletic facilities—locker rooms and gymnasiums being major users of heat for outside air—arrange the equipment so it will recirculate room air during those periods of non-occupancy. Most locker room equipment handles 100% outside air according to code requirements and usually does not have the ability to provide recirculation. During basketball games, try to re-use the heat generated when the people, the players, and the lights keep the temperature elevated in the gym.

If you have an indoor swimming pool, provide a good filter to help reduce the amount of make-up water and also provide a reduction in fuel required for heating this water. The exhaust and supply system to the indoor pool buildings should be regulated by dew point controls so that moisture does not condense on the structure and deteriorate it. Such controls should also reduce the amount of supply and exhaust of outside air which would help considerably in reducing the amount of heat input required to keep the pool space warm and usable. If the pool if not too large and it is possible, provide a cover for the pool water level itself so that the moisture does not evaporate into the air and require de-humidification.

In auditoriums and theaters, make sure the air distribution is such that it gets to the back of the theater, does not disturb the stage curtain, and still does not create a high noise level. Also, provide a control that anticipates filling up the auditorium so that there is no temperature lag between the time people come in and the time the program starts. We have measured time cycles in a number of auditoriums and find that it takes about 15 minutes for the thermostat to react and cool air to enter the seating space after a program has started.

We have been successful in theaters as large as 4,400 seats, using a step ceiling with adjustable grills carrying the air clear to the back and still not creating a high noise level. In this case, the return air was vented through the floor into a ceiling space in a convention hall below, using a number of floor outlets. Be very careful under balconies so that the air distribution does not cause drafts or does not become warm because the air from overhead did not get to the space. Don't forget the orchestra pit; it needs cooling too.

In auditoriums and in theaters, both legitimate and movie-type, if you have the personnel or the computer facilities, limit the amount of outside air to the population in the space. I have been in 500-seat theaters with five people in the audience and in 2,000-seat auditoriums with 500 people.

If you have outside air intakes below ground where you have a pit in front of the louvers, it is suggested that you put a solid wall at least waist high around this pit so that the dirt from the ground and other materials do not blow into the pit, creating additional contamination (see Figure 15-17).

We have successfully used cooling towers with sound absorbers in a pit with high discharge velocity (see Figure 15-18).

In residences, our biggest problem is the fact that they quite often use flexible fiberglass duct, and try to fish it through the framing of the roof. In many cases, this pinches off the duct work and creates a resistance. When designing the duct work, provide it big enough so the static pressure in the entire system does not

Figure 15-17. Outside air intake below ground.

Figure 15-18. Cooling tower with high velocity sound absorber below ground.

exceed .15" S.P. so that you can get a reasonable air velocity to the air outlets. On high ceilings, use an air outlet that projects the cool air to the floor rather than trying to spread it out at the ceiling, hoping it gets down to the bottom; usually it does not. If you really want to do a good job on high ceilings, use a reversible ceiling fan, run it in reverse in winter and directly down in summer. Floor-to-ceiling windows seem to be popular today and if they are on any side that the sun is shining, you have to blow cool air over the face of this glass to have a reasonable temperature within the room. This is true both summer and winter. If the air-handling unit is in the garage, make double sure it is air-tight so that it does not pull car exhaust fumes into the house.

In boiler plants, if you use an economizer on the stack, make sure there is sufficient draft to take care of the boilers. We tested one plant of eight boilers and the system worked fine on six boilers, but when the last two boilers came on there was insufficient draft and the fire blew out the doors. It was necessary to put a bypass around the economizer with a motorized damper so that when more than six boilers were in operation, the damper opened up to have sufficient draft on the system.

Special rooms for smoking are beginning to be of interest in some companies because of the restrictions on smoking. Such special areas need a good system of filtration to remove contaminated air, and a good system of ventilation when required. Electrostatic filters can take most of the smaller particles out but require a higher static pressure. You also should have a means of flushing out the space with 100% outside air so that you will not change the humidity level too much in the space during this flush-out period. The space should be on slight negative pressure so the poor-quality air does not get into the other occupied space.

On smoke removal systems, provide either back draft dampers or motorized dampers so that at night when you are shutting down the system you will not have outside air penetrating the building and decreasing or increasing the humidity.

CHAPTER 16

OPERATION SIMPLIFICATION

If you can afford it, prepare an actual operation manual so that the owner and the operator can find out how the system is supposed to work, what temperatures and conditions are required, and what to do if the system is not functioning properly. Contractor-prepared maintenance and operating manuals usually consist of suggestions from the manufacturers on maintenance of particular items. In most cases, the contractor himself does not know what to tell the owner, so it is up to the engineer who designed the system to put down on paper some information as to what the system is supposed to do, how it does it, and how to keep it going. The manual should also be in understandable language, not like computer manuals which are designed for specialists. If you are using a computerized control system, put the information on the computer screen with the controls—we will discussed this later on. Tell the operator how to adjust an air outlet and under what conditions he should make these adjustments. (See reference book).

An example of management system training program failure occurred with two buildings totaling 300,000 sf and operated from a single control panel. The system had over 900 thermostats and control devices. Thirteen years after the buildings were built and occupied, we were called in to find out why their utility bills had doubled from the time they started occupancy. Investigation revealed there had been six or seven different operators of the system. No information was transferred from the first one—who was trained—to any of the others. When we checked the system, we found that the operator was only turning the switch on when he felt like it, when there was a complaint, or when he felt that the system should be shut off on Friday night and turned on on Monday morning. There was provision for start-up of part of the system in case it was occupied on weekends. This was a hot- and cold-duct system and as you see from Figure 16-1. The control temperature varied widely since the operator was constantly changing the deck temperatures to catch up with the system. This failure in training and operation was wasting almost a half million dollars a year in utility costs.

There are now available some small-size outside air supply units specially designed to provide relatively low dew point and high latent removal for such an application. There are now some

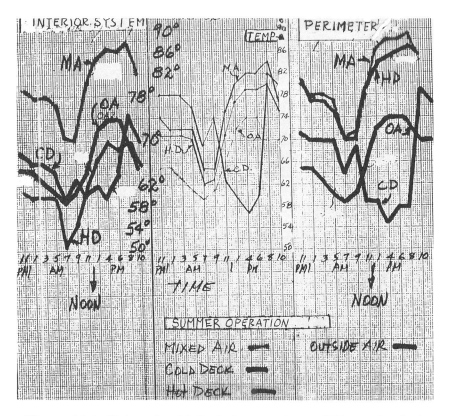

Figure 16-1. Hot and cold deck temperatures. MA = mixed air OA = outside air; HD = hot deck; CD = cold deck.

systems that are using floor-mounted air outlets which may have some merit if the velocity is not too high and the air is not too cold. If there is low velocity and cool air, the air may lie on the floor and not be distributed to the areas where you really need it. In many offices which are broken up with partitions, the air supply should be such that air gets into each of those small areas, without drafts. The usual design of an office building is with individual offices on the outside walls and a large area with partial partitions between work stations in the center. Air distribution from side wall grilles is not satisfactory and there should be good ceiling supply area and good ceiling patterns in order to get the air where it is most needed. There also should be provisions for different controlled air at different temperatures to those parts of the building which are exposed to sunlight, even though the glass may be solar absorption type.

If you have an office building which has areas occupied nights and weekends such as a club or a restaurant, make sure that you provide a separate air conditioning system and a separate refrigeration plant for those areas. We had one building of 180,000 sf which had a club room on the top floor which operated every night and on Saturdays and Sundays. Because the building had 200-ton chillers in it, it was necessary to operate one of the chillers at 177 tons to get 56 tons in the club room because that is as low as the chiller would operate.

If you are working in a large computer area, that requires an uninterruptible power supply, consider using a gas engine or diesel generated plant for that purpose and using the waste heat perhaps for an adjoining office building for both heating and cooling, using absorption chillers. We analyzed a 2,000 kW computer center and found that we could provide a generating plant at less cost than a battery powered UPS and provide both heating and cooling for a 100,000 sf office building in the upper Midwest.

If the installation is lower temperature like refrigerated warehouses, consider the use of more than one large compressor since the conditions of operation may vary. We inspected a warehouse that was designed for 30° and was operating at 65° since it had been changed to storage of paper materials and the very large compressor was shortcycling on a 5-minute cycle which not only was damaging to the system, but almost doubled the demand charge on the electric bill.

In refrigerator and refrigerated display cases such as those in supermarkets, do not supply any air to those areas as the air motion has a tendency to remove the cooling from the case and also cooling is not required because of the radiation from the cooling cabinets. One place in supermarkets which should have more air is the check-out area where there is a concentration of people and where people are coming into the store, to provide greater comfort.

If you are using an air door either wide open or with plastic strips, try to make sure that there is no positive or negative pressure at that location or you will lose considerable cooling or heating. This is particularly true on those buildings which have truck openings for delivery. Unless the receiving and shipping areas are tightly closed, warm air or cold air gets into the building rapidly, increases operating costs, and causes difficulty in temperature in those general areas. We have measured air velocities through high-performing slot air curtains at wind velocities over 20 mph with the negative building pressures drawing velocities as high as 2,600' per minute through openings.

In high-temperature areas such as show windows, especially on the sunny sides of east, south, and west, instead of trying to blow cold air onto those areas, try exhausting the air or returning the air from those areas and let the air in the building slowly flow towards those sections and indirectly cool them. This is also true for storage areas like in department stores and other retail stores.

If there is an attic or a confined space above the ceiling where the duct work is running, see if there are provisions for inspecting the duct work at a later date to check for leaking. We have tested a number of systems 5 to 10 years old where not only the fiberglass joints and tape were loosened up and losing about 30% air, but also sheet metal duct work which was losing air at the slip joints. In one case the sheet metal duct was losing 30% of the air supply.

Make sure that the building insulation, if used, has a vapor seal on the inside in the northern part of the country and on the outside in the southern part of the country to control moisture penetration through the walls and attic ceilings. If improperly installed, this may cause high or low humidities in the building and you may be blamed for it because your air conditioning system did not take care of it.

In the industrial field, the pattern of ventilation and exhaust is very important. If there are areas which have very high exhaust, you should supply outside air to those particular areas rather than try to pull it through the entire building. Many times objectionable materials, such as plating and fluxing chemicals, are picked up and transferred to another department and cause damage. If there is a lot of heat generated by equipment on the floor, try to take the heat off the ceiling and put it back towards the floor in winter. If you are using radiant heat, be aware of the locations of materials used in the area. Metal products may absorb heat, leaving personnel in the area too cool.

If there are dangerous chemicals in the area, specify some sensing devices to alarm, should they get into the atmosphere. If there are particulates of any size in the air, or oil vapors which may not be harmful to personnel, and you are using cooling or heating coils, specify not more than 8 fins per inch on the coils or they will plug up rather rapidly. Also, do not use more than 4 row tubes in a section and, if more tubes are required, space them so you can clean in between. Here, again, use back draft dampers and outside air dampers on supply and exhaust equipment so the building can be sealed up at night or during non-occupied periods to keep the outside air out of the building and lower the energy costs.

If you have a pumping system of any kind and you require low circulation under certain conditions, don't try to use your big pumps and slow them down since, in most cases, they will not provide the discharge pressure you need. Put in an auxiliary small-size pump with the proper check valves to operate under low-circulation conditions.

If you have an industrial process which is using chilled water and also have an air conditioned plant, investigate the possibility of isolating the two refrigeration systems since the process system may be operating at a lower water temperature. Your operating costs will improve with a separate plant. We experienced this in a plant making tennis shoes, where the manufacturing equipment required 35° water and the air conditioning system required 45° water. Isolating the two systems greatly reduced the operating costs.

On medium-pressure air systems, be very careful with the location and the amount of elbows and turns to keep the static pressure within range. We have experienced pressure drops as much as 3" across two elbows, even with turning vanes in a 4" static pressure system.

On test and balance reports, specify that the test equipment should be calibrated within the last 6 months and that the reading should come within at least 10% of the specified amounts and examine the reports carefully to make sure they are somewhere within the range of what you specified. For instance, a hospital air conditioning unit supplying patients' rooms was specified at 16,000 cfm, actually tested out at 13,500 cfm, and 10 years later was tested at only 10,200 cfm. Under those conditions the system did not operate at the leaving temperature specified and had to be modified. Quite often we have measured air outlet capacities considerably different than those specified and that reported by the contractor. It might be a good idea if the engineer would spot check air velocities and temperatures after the system is balanced to make sure the contractor performed properly.

Most engineers specify a certain number of hours or days to instruct the operator of the system but do not mention the qualifications of the person doing the training or what he's supposed to perform in the training period. Training in energy management, how the system works, and what to look for in case the system is not performing is important. Tests should be made under both cooling and heating conditions to make sure the system operates properly. This is critical because a system is turned over in only one weather period. The contractor should come back and retest the system in the other weather period and make whatever corrections are necessary, to ensure correct system payoff figures.

Sometimes it is a good idea to check up on the installation. We had designed a ceiling-mounted unit for the blood bank area in a hospital. We got a call that the temperature was rising in the space and was endangering the quality of the blood. We ran a test on the air conditioning system and found that the contractor had set the fan speed on low rather than on high which was necessary for the room. We corrected it just by changing the fan speed on the 3-speed motor.

On the sunny side of the building, the shades should be drawn to prevent radiant heating entering the space, especially if it is a small office and the occupant has his back to the window.

A 17-story apartment house had two sections. There were two apartments on every floor of each section, and the two sections had separate elevator lobbies. These lobbies had 4,000 cfm of outside air being pumped into them, when only about 1,000 cfm was required. The excess cost was around $10,000 annually.

CHAPTER 17

CONTROLS

There is a lot of good control equipment on the market, but much of it is improperly applied, in poor locations, and with little or no instructions on how to use it.

Putting thermostat in the return air wall on an equipment room keeps them from being tampered with, but it is much better to use a remote sensor and put the stat in the room where nobody can touch it, but where it will still sense the proper temperature in the space.

Even remote control can sometimes cause trouble. We were in a theater recently where the temperature got quite cold simply because there was a flood light shining on the thermostat during a lecture period. Someone went over to the manager and asked him to correct the situation. He said, "I'm sorry; even though this theater is in Florida, it's controlled from Texas."

The two best methods of saving energy with controls are by shutting the system down at night and on weekends, and by controlling the start-up of motors if there are multiple units. Changing the control temperatures between occupied and unoccupied periods also helps.

In many cases, problems encountered during occupancy are due to the fact that thermostat are readjusted sometimes by occupants pushing the control to the maximum low or high limit. If you can get a thermostat that has limits on how low and how high it can be set, preferably within a 4° range, you will save your occupants a lot of trouble.

Specify that the thermostats should have thermometers on them and tell the manufacturer to make sure that the thermometer and the thermostat are properly calibrated. We have found thermostats to be off as much as 7°. I have previously mentioned such problems with controls on dishwashers, kitchen exhaust hoods, cross-connecting zones, and special controls for continuous cooling in areas like computer rooms.

If you are going to specify a computer management system, add a description of how the system works to your specifications rather than just the technical data of the characteristics of equipment, the manufacturer, and how it functions (see Figure 17-1 and 17-2).

Normally the specifications mean nothing to the owner. He doesn't know what he's getting, so if you would describe what the system is supposed to do, as shown in Figure 17-3, the owner might understand what he's getting and he may want to make some modifications to it if he feels that he needs more control or has too much.

Energy-management systems software should be designed with a computer screen that is understandable by lay persons rather than only by computer technicians. For instance, why not put a simple diagram of an air conditioning system on the screen with the list of the sensing points like outside air, return air, supply temperatures, static pressures across filters and across the air conditioning unit, room temperature, and room humidity? With air-cooled condensers, use a leaving-temperature sensor to determine if there is recirculation or if there is any interference with the system. Include also on that screen the original design and test conditions that the system should work at, showing the numbers in green and also the actual operating conditions showing the numbers in red if they are too high, and blue if they are too cold.

I have mentioned conditions where cooling towers and air-cooled condensers are interfered with and the efficiency dropped, but the temperature across an air-cooled condenser can also determine the conditions of the refrigerant circuit. An example of this was a system which was only 3 years old where the temperature rise on the air condenser was 30° instead of the normal 20°. I checked with the operator and found that the equipment had been

General

Each control panel shall be capable of full operation either as a completely independent unit or as a part of the building-wide control system. All units shall contain the necessary equipment for direct interface to the sensors and actuators connected to them.

Control strategies shall be owner-definable at each control unit, and for all the control units in the system from any one operator terminal. Each control unit shall provide the ability to support its own operator terminal if so desired.

Each EMS panel shall include its own microprocessor, power supply, input/output modules, termination modules and battery. The battery shall be self-charging and be capable of supporting all memory within the control unit if the commercial power to the unit is interrupted or lost for a minimum of 45 days.

Building Control Functions

Each EMS panel within the building control system shall perform both temperature control functions and energy management routines as defined by the operator.

All temperature control functions shall be executed within the EMS panel. The user shall be able to customize control strategies and sequences of control, and shall be able to define appropriate control loop algorithms and choose the optimum loop parameters for loop control.

On/Off Outputs

Control panel shall internally provide test points for the circuit driving the equipment contactor, for the purpose of troubleshooting the 24 VDC circuit to the contactor. All such relays or digital output modules shall provide a pilot light or LED display of this same status.

Modulating Outputs

Modulating outputs shall be industry standard 0-5 VDC, or 0-12 VDC with definable output spans, to adapt to industry available control products. Milliamp outputs of 0-20 mA or 4-20 mA are also acceptable. Drive open/Drive closed type modulating outputs are acceptable provided that they also comply with the following requirements.

Figure 17-1. Usual control specification 1.

5.0   SEQUENCES OF OPERATION

5.1   TEMPERATURE CONTROL SYSTEMS

5.1.1   Variable Air Volume Air Handling Unit

5.1.1.1   Timed On/Off Control
The SDC serving the particular unit, provided that the fire alarm is in the normally closed position, shall start the fan. Status of the unit shall be verified from the SDC or the existing CHS through a differential pressure switch located across the fan.

5.1.1.2   Temperature Control
The SDC shall stage the different steps of refrigeration in response to the desired setpoint and based on the supply air temperature.

5.1.1.3   Pressure Control
The static pressure sensor located approximately 2/3 down the duct run shall provide the SDC with an electronic signal directly proportional to the duct's pressure. The SDC shall modulate the inlet guide vanes located at the fan in response to the static pressure in order to maintain the desired pressure in the duct.

5.1.1.4   Monitoring Control Points
Return air temperature and filter status are provided at the SDC for monitoring purposes and in order to aid the operator in the troubleshooting process.

5.1.2   Exhaust Fans

5.1.2.1   Timed On/Off Control
The SDC shall start and stop the exhaust fan according to a time clock set by the owner at the host computer.

Figure 17-2. Usual control specification 2.

Description of Operation of Energy Management System

1. Computer will start up all air conditioning units (18) in steps at a predetermined time in the morning. Computer room air conditioner will not be shut down at night. Air conditioning units with two compressors shall also be staged to control electric demand.

2. Air conditioning unit fans will run continuously during occupied period.

3. Room temperature sensors will operate compressors in steps on air conditioning unit supplying each respective area.

4. Computer will display room temperature for each zone.

5. If existing room temperature is unsatisfactory, temperature setting of sensor may be reset at computer.

6. Computer will indicate if compressors are running.

7. Humidity sensors in library and warehouse will override temperature sensors in those areas to control at preset relative humidity (maximum 60%). Also restart at night and weekends.

8. Major computer room temperature sensor shall automatically switch to standby air conditioner if temperature in room rises to 4" above sensor setting.

9. At each room temperature sensor a manual switch is provided to restart area air conditioning unit for a period of three hours, if the system is on unoccupied cycle.

10. At a predetermined time, all air conditioning units, except computer room, shall be shut down for unoccupied period at night.

11. For weekends and holidays, computer will automatically start up all air conditioners in steps for 2 hours per day to prevent humidity buildup.

12. For break room, computer can open outside air damper, if desired, to flush out room with 100% outside air.

13. In conference/training room, a manual 3-hour timed control will run air conditioner at high speed for high-occupancy periods. At low occupancy, fan will run at low speed to reduce electric use.

14. Parking lot lights will be controlled to turn on and off according to need.

15. Water heaters will be shut off nights and weekends.

16. Energy management system can also be controlled from a computer at home through a modem line, as well as the computer on site.

Figure 17-3. Understandable control operation.

cleaned a few weeks before by the air conditioning contractor who removed the cooling coil, cleaned it, and then reinstalled it. He forgot to totally evacuate the refrigerant circuit and completely fill the system with refrigerant. It was partially filled with air and was causing a higher temperature.

If there is unoccupied space in the building for extended periods of time such as empty office space or empty hotel rooms, provide a control that will shut down the air conditioning to those areas when not used. In high-humidity areas, turn the system on at full capacity for those particular rooms for 2 hours a day and the mold and mildew contamination will be taken care of. Water pressure sensors at the top of a high-rise building are very important. We found a number of cases where there was insufficient domestic water pressure to get to the top of the building with enough residual pressure to operate equipment that was up there. Many city water systems do not have sufficient pressure to raise the water that far, and booster pumps are required.

With proper information on the screen, maintenance of the system may be reduced. The operator may determine the problems at the screen without going to the equipment to make individual measurements. This is also true on multiple building systems where a maintenance man would have to travel a distance to get to the equipment to check its operation.

If you can't work out these instructions to put on the screen, you might refer to a manual listed in the reference called "Analyzing Field Measurements of Air Conditioning and Heating." A real need for computer input was shown with a 40-ton air-cooled unit which was using nearly twice the electrical input for the load. The owner complained to the power company that his meter was wrong. They replaced the meter and it still read the same electrical usage. Tests on the unit determined that the bypass valves on the compressor were stuck in the open position, and that the 40-ton unit was delivering only 5 tons at 25 HP. If there had been remote indicators on the amps to the motor, and inlet and outlet water temperatures on the chiller, this could easily have been picked up.

Remember the maintenance man operating the computer has other things to do besides sit in front of the screen. If he has color coding, he can glance at the screen and determine if most of the system is working satisfactorily, without spending a lot of time punching buttons.

Another procedure which may help on VAV systems is changing the supply air temperature when a considerable amount of air is reduced on the system. A lot of units have bypass ducts between the supply and return so that the volume of air can be reduced without fancy speed control or volume dampers on the fans themselves. It may be a good idea to provide a sensor in the bypass which measures velocity so that when it reaches a certain point, say 50%, the supply temperature can be raised approximately 2° which may re-adjust the dampers in most of the system keeping more air motion—which is what you want—and still control the system properly.

In room cabinet units with self-contained controls which say C-H, install built-in thermostats for low and high points which are not adjustable. This prevents waste heat and over cooling which cause mold and mildew in high-humidity areas (see Figure 17-4).

If you have a return air fan in your system, make sure it is properly interlocked with the supply fan so that it does not exceed the supply fan capacity. We tested a building which had a variable speed motor on the supply fan but a constant speed on the return fan, and when we measured the outside air supply we found that the air was going *out* the louver rather than entering it.

If you are using time clocks to operate motors such as shutting them down at night and on weekends, don't use a standard time clock for more than perhaps two motors. The motors should be staggered because it takes about 200% inrush current to start a motor. Continuing at this rate for an extended period of time will increase the electrical demand and raise costs. The standard time clock it has a minimum space of 15 minutes. If you have a lot of motors and use this type of time clock it may take as much as an hour or two to start up all the motors (see Figure 17-5).

Figure 17-4. Cabinet unit control.

If you have an electronic type time clock, make sure it has an automatic correction for power outages, so that if you are on an off-peak rate you don't get put back on the standard rate. We experienced this in a country club clubhouse where the golf carts were being recharged on an off-peak rate. Recharging 100 carts cost about $1,150 per month with the on-peak rate and $430 per month off-peak. This particular electronic time clock had a series of instructions, as indicated in Figure 17-6, so that it was *manually* corrected for power outages. The first month it was in operation somebody forgot to check it one morning with the result that it went on-peak during the night for about an hour. The electric bill went back up to the normal rate and cost them nearly $600 more that month.

Step start-up of motors is very important in any type of building. In one building with 16 air conditioning units the de-

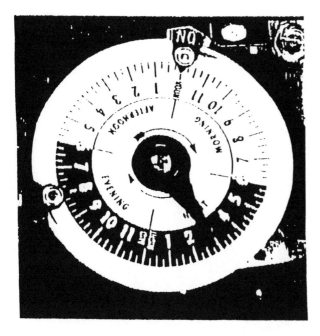

24 Time control
15 Minute intervals

7 Daytime controls
15 Minute inter-
vals
Difficult to adjust

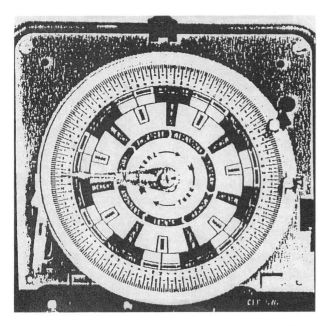

Figure 17-5. Time clock.

Electronic Time Clock Instructions

1. It is imperative that, the first thing every morning, someone looks at the screen to determine if there was a power outage overnight.

2. If the screen is flashing, please do the following:
    a. Press Exit. Screen will show amount of time of power outage.
    b. Press Exit again and screen will display the correct day, time and date.

3. To preview master entries:
    a. Press Review
    b. Press Program
    c. Press 0
    d. Press Exit

4. Manual override
    (turn on)
    a. Press Manual
    b. Press 1:2:3:4: (you must use colons
    c. Press Next
        (turn off)
    d. Press 1:2:3:4: (you must use colons)
    e. Press Next

Figure 17-6. Electronic time clock instructions.

mand dropped from 360 to 240 kW, considerably reducing their electric bill. Another important use of timing operation is during cleaning periods and janitor service in office buildings, particularly. We tested a 300,000 sf office building where the cleaning people ran the air conditioning full capacity during those periods. The cost of this cooling was $200 000 per year. Tests were made by shutting down the air conditioning system on Friday afternoon, and leaving it off entirely until Monday morning. With an average outside temperature of 85°, it was discovered that the temperature only rose 2° in the building and no cooling was required. So make this provision to shut the cooling off during these periods and let the building coast.

Another good place to control exhaust is in kitchen hoods where cooking is not done all day long. Provide some means of automatically shutting off the exhaust when not being used but provide a manual switch with a timer alongside the hood in case the chef wants to come back and use it for a while.

Nearly all dishwasher hoods are designed improperly. The hood should consist of two sections: one approximately 2' deep covering the entering side and one 2' deep covering the leaving side for the most efficiency. Also shut the fan off when the dishwasher is not in use.

Tell the contractor to furnish a good digital thermometer— portable type—to the building operator. It is amazing the psychological effect on a tenant when in case of a complaint he can take this thermometer and show it to the tenant. We solved this problem in one office building where the control system was not working too well and the maintenance man had to constantly adjust the air outlets to satisfy his tenants. I bought him a digital thermometer (this particular one would flash the numbers) and he showed it to the tenant and solved the problem. He's back doing what he's supposed to do.

As I previously mentioned, gauges and thermometers are very important. For instance on the suction side of fuel oil pumps from oil tanks thermometers and pressure gauges can indicate whether the fuel oil is too cold or viscous for the pumps to provide the proper amount of oil to the burner which may result in low efficiency. On condenser water systems, thermometers will indicate whether the cooling tower is working properly, and show failures in the condenser system (especially in heat pumps).

If you are specifying an energy management system in a luxury apartment house or condominium, it may be a good idea to add to the system some security alarms, including a moisture sensor on the floor of the air conditioning equipment room and a high-limit sensor in the apartment space next to the thermostat. Also provide an alarm panel at the security desk. We have experienced very high costs of repairs and replacement due to flooding, sewer back-ups, and organic deposits. You might also add an alarm on the elevator in case it is not working. This would not HVAC oriented, but is a good idea. In an enclosed garage, a carbon monoxide sensor wired to that same panel would also be helpful. If there is a central chilled water plant and boilers, the required alarms for refrigerant leaks and for carbon monoxide from flue gases should be fed back into that same security panel.

If the building is of any size and has an energy management system and security it is a good idea to put an alarm device either lighting or bell at the security office. If anything in the HVAC system goes far off of normal operation then the security guard can call the maintenance chief who should have a lap-top at home tied into the same system. He could then determine where the problem is and perhaps solve it without even going back to the building.

CHAPTER 18

AIR POLLUTION

Today air pollution can become more important than energy reduction when it results in sick personnel and lost man-hours. Law suits for health problems can be costly as well.

Our experience indicates that the major problem is air circulation. Air motion over your body is necessary to remove heat and moisture for summertime comfort. It is necessary to provide a warm envelope around your body for winter comfort. And the air should have a high enough humidity to not to remove too much moisture from your body. Air quality and circulation are key elements in keeping people relatively comfortable.

The VAV system is the biggest culprit in this condition because of the low air circulation and the change in the air pattern from the air outlet to the space, caused by the reduction of air volume and the change of the characteristics of the air outlet. If you must use a VAV system, use a fan VAV box at those areas where this particular condition may occur so that there is constant circulation within the space.

In one project we used fan VAV boxes instead of reheat in a system which eventually saved nearly $2 million a year for an investment of less than $4 million. As I have already described, temperature and humidity levels are very important even though different people require different conditions to make them comfortable. ASHRAE indicates that the effective temperature or comfortable level varies between about 70° and 76° and 30% to 60% relative humidity. In dry areas it takes a little work to get up to 30% to 40% and in high-humidity areas it is sometimes extremely difficult to get below 60%. Outside air should be supplied to the building through an air conditioning unit. It should be either mixed with return air or, if supplied directly to a return air space (such as a dropped ceiling in a return air space), the air should be conditioned close to the temperature of the space to prevent high dehumidification in dry areas and mold and mildew in high-humidity areas.

The balance of air pressure within a building should be on the slight positive side rather than on the negative side to prevent contaminants from coming in from the outside through doorways and openings. In the case of high-humidity areas, slight positive pressure reverses the flow of air through walls and reduces the contamination of insulation.

If there are odors in the building, and you want to flush them out by pouring 100% outside air into the building for certain periods of time, make sure the temperature of the duct is fairly close to the air temperature to avoid condensation and mold growth.

If there is an enclosed garage on the building, create a negative pressure there to keep carbon monoxide fumes from cars out of the building. You might put in a carbon monoxide defector to turn on a fan, unless you want to run it continuously, to make sure this level does not get too high.

Make sure the exhaust from the laundry room is nowhere near any operating equipment. We tested a motel operation where the discharge from the laundry was right next door to the intake to an indoor cooling tower. The cooling tower was completely plugged up with lint.

Make sure that there is a slight negative pressure in kitchens so that odors and contamination do not float into the dining room. Do not make the negative pressure too strong or it will draw too much cooling out of the dining room.

If a building adjoins a major highway, such as an interstate, place the outside air intake as far as possible from this area, as you may get fumes from the passing traffic into the air stream.

In hospitals, especially, make sure that the exhaust system is isolated from toilet rooms and isolation areas so that the air removed does not mix with the exhaust going through a heat recovery unit if such is installed. This is particularly true of heat wheels because the contamination can be picked up in the wheel and put back in the air supply, even though there are hepa filters in the system. All the contaminates may not be removed.

Be careful to avoid noise pollution. Put equipment on vibration isolators if there is a possibility the unit vibration may penetrate to another area. The same is true of flexible connections from the unit to duct work. Many noise pollution problems can be solved by using sound-absorbing materials in equipment rooms, making sure the materials do not absorb moisture and cause bacteria growth. If insulating duct liner is used on sheet metal duct provide inspection doors or openings near turning vanes where used. Then inspection can be made at a later date to find out if a liner loosened up and partially plugged the turning vanes.

We have been disappointed in the use of so-called electronic filters which use air flow to create ionization on particles. We have been more successful with high-voltage ionizing units with opposite charge collection plates which can be removed and washed. There are also some new felt and fiber filters on the market with extended surface which seem to be much better than the old fiberglass type.

Make sure that you specify access so that drain pans can be completely cleaned periodically, since quite often they are coated with materials which do not always clear out. We have inspected drain pans with a so-called chemical pad to eliminate bacteria and hardness growth and found them unsatisfactory.

Sometimes minor contamination can cause difficulty. In a department store on a newly installed system, a ceiling got dirty around the air outlets in one area and appeared to be clean in another area. Examination of the dirt showed that it was lint from the carpeting which was dark in the dirty area and gold color in the light area. We have also had problems with dust on an executive's desk with a lay-in, perforated ceiling above.

Again, examination under a microscope indicated fiberglass particles. Examination of the ceiling panels showed that the insulation which had been installed in a paper bag had deteriorated and was falling through the ceiling.

We cannot overemphasize the importance of keeping moisture and contaminants out of cooling tower intakes. Considerable amounts of bacteria and contaminants have been found in cooling towers, including Legionnaires Disease bacteria. The requirement for either high velocity discharge from cooling towers or arranging air intakes so that they will not get this air is extremely important.

In certain industrial processes where you are exhausting dusty materials know what these materials are. A ceramics plant had a series of bins of different materials which were picked up, put into a container on a car, and then run through to the processing plant. It happened the dusty materials which were exhausted were the least injurious and the materials which were not dusty but had no exhaust were those which would cause health problems.

Interior materials today, except perhaps new carpeting, give off very little of what we call "off gases" which may interfere with the health of occupants. It might be a good idea to provide some control so that the entire building can be flushed out with outside air at night or whenever outside conditions are favorable to clear any odors and contaminants. Make sure there is some provision for keeping excess moisture out of the building (which will cause organic growth) and excess dry air (which will require considerable humidification.

Specify minimum air supply temperatures such as approximately 55° for most applications, unless you have special ceiling outlets. We have run into mold growth on air outlets in high-humidity areas and on drafts in low-humidity areas. The same is true for heating temperature. Keep supply air temperature at not much over 100° (120° tops) so you will not get uncomfortable blasts of warm air.

If you are using hot water heating it is a good idea to vary the water temperature in proportion to the outside air so that you get more continuous circulation of water, rather than trying to reduce the water flow with a variable speed pump. You will get better quality and better comfort to the occupants.

Owners should inform architects where they plan to install copying machines and similar devices which give off ozone and other contaminants that are injurious or irritating to personnel. In the case of copying machines, provide an exhaust grill near the machine with a positive fan to blow the air up into the return air (especially if it is a ceiling space) and let it dilute into the main air stream.

If there is an incinerator on the property, such as is normal for hospitals make sure the flue gases do not get into the outside air intake. We had a case where nurses in a hospital were getting headaches and were dizzy. We found that the wind was blowing the flue gases from the heat recovery boiler on the incinerator into the outside air intake. It happened that they were burning urethane bed pads in the incinerator which was giving off phosgene and a little bit of cyanide.

A space heater vent failed when the heater was installed with a vent going outside the wall using a dryer vent tube with a back draft damper. Two people were killed in this case. The manufacturer of the heater was sued and was required to pay. Specify that proper venting be provided in all cases of combustion equipment. I was chairman of the building commission in a million-population city for 28 years and one year we lost nearly 35 people due to poor ventilation from space heaters. We also lost (nationwide) over 30 maintenance people who worked on air conditioning units basements where Freon floated over to a gas water heaters. The flames in the water heaters broke down the Freon to phosgene, a deadly gas. To my knowledge, the manufacturer of the refrigerant was never sued even though they did not put any information on the product stating that it could break down under those conditions.

For an existing building which is partially remodeled and it is partially occupied during construction, specify the requirements for isolating the air handling system to the occupied areas to make sure that dust, debris, and contamination picked up during construction do not get into air supplied to the occupied area.

CHAPTER 19

CORROSION

Corrosion causes a lot of difficulty in equipment. I have mentioned several cases, but atmosphere and water and chemicals can cause corrosion after a short period of time. Here are some examples.

A cooling tower with an aluminum hub on the fan failed after the first year, since the water was being taken from a river and it was so high in sulfur that it ate out the aluminum hub.

A restaurant dishwasher had a stainless-steel hood which was deteriorating because of the type of detergent used.

Near the ocean or any saltwater area, salt is distributed in the atmosphere and can cause serious corrosion to aluminum fins on coils. Some coatings are satisfactory and some are not. I inspected a 3-year-old, 10-ton air cooled condensing unit which had an epoxy coating on the fins. The epoxy coating had melted, and run down to the bottom of the coil, plugging up about 1/3 of the face.

Radiant heating systems are subject to corrosion. For instance, in a garage, moisture crawling through cracks in the wintertime rusted the pipe on the outside because of the salt in the moisture dripping from the cars. We have seen corrosion on the outside of piping which was insulated with rigid type material, but not properly sealed on the outside. The surfaces of the pipe on chilled water was wet all the time and it rusted from the outside in. This can also happen in large cooling towers made of metal which deteriorate because of poor atmosphere or water treatment. We had a cooling tower only 4 years old in Florida where the outside louvers had completely rusted through because of poor water treatment. We had a cooling tower in Hong Kong which we had to replace with a wood structure because of atmospheric corrosion.

Electrolysis, of course, causes great difficulty in heat exchangers and in boilers where tubes may corrode due to electrical circuits caused by opposite metals or electrical grounding. We had a boiler in which the tubes rusted out in 3 years because the ground of the electrical system was to a plastic water pipe.

Underground piping, of course, is subject to corrosion, depending on the surrounding moisture content and chemicals in the soil. If you are using underground piping, and it is insulated, make sure that the jacket is watertight and impervious to any corrosive materials that you might find in the area.

Water treatment systems are extremely important because nearly all water has some minerals in it. We tested a 4-year-old office building which had a water treatment system in the cooling tower circuit, but it had not been properly maintained. The cooling tower louvers had practically all rusted out, with the cooling tower half plugged up. The piping system in this 16-story building was partially plugged up with lime. Since the building was not entirely occupied and the heat pump units had not been installed in about 5 floors, sections of the piping were removed for inspection. It was found that on at least 3 floors the piping was entirely plugged up and there were pin holes in the piping that were leaking.

In a group of apartments in a number of buildings with underground piping, the shower controls were not working because they were plugged up with lime. Examination of the system indicated that the city water was coming in, going through a hot water boiler—good sized one—then circulating to the various buildings underground without insulation. This problem was solved by removing the control unit from each shower valve, replacing it, and then cleaning the original unit by dipping it in a lime-removing solution. Also, to prevent the boiler from liming up, a heat exchanger was installed using circulating water from the boiler into the heat exchanger. A separate circulating pump was used to pump the domestic water through the heat exchanger to the various apartments.

CHAPTER 20

COURT CASES

I have been an expert witness, and sometimes a defendant, in quite a number of court cases. The following stories may be of value to you if you find yourself in this position. You can't always tell the results, regardless of the technical accuracy of your statements because of the fact that most juries are not technically inclined and must have all the information explained in layman's language. To give you an example, when talking about the voltage and the power of an electrical circuit, I use the description of a garden hose saying that with low voltage, the water just drips out the end of the nozzle, and at high voltage the water squirts out 20'.

Sometimes you don't win. I was a witness for the failure of a radiant heating system in a large home. The heat piping had been sheet tubing and had not been provided with the proper loop expansion joints in the slab. The system failed in three years, and the owner had to remove all the contents of the house so the floor could be chopped up and the heat piping replaced with copper tubing. The contractor sent the owner a bill for $1,600 which the owner refused to pay. The contractor placed a lien on the property and the court case was to remove the lien. I was called to testify as to the quality of the installation. I examined the pieces of concrete with the tubing in them and testified.

After I had testified, the attorney I was working with said "You won the case." About three weeks later, I saw the heating contractor (who I knew) on the street but, of course, couldn't recognize in court. I asked him what happened to the case. He said, "You lost it." I said, "How?." He said "Well, Monday morning the contractor put on his overalls and took the jury out to this house. Out in front of the house were two European sports cars, inside the house they saw two empty whiskey bottles in the sink, walked into the den and there were two nude pictures on the wall, they went back to court and found against him (the owner)."

In another case where a crane had accidentally touched a 13,000-volt line, a man standing on the ground leaning on the tread was killed. The operator inside the crane cab was not hurt. The family of the deceased sued the power company and collected, even though the power company had analyzed the overhead cable and determined that the metal on the cable compared with the metal in the cable on the crane.

In another case, a man was putting up an outside TV antenna and touched a wet tree. The current went through him, burned his shoes off and, of course, killed him. Again, the family sued the power company. It took a long time, but they found that the insulation on the power line had rubbed off and the line had grounded through the wet tree.

In another case, a contractor sued the architect and the engineer for improper heating design in a one-story dormitory building at a university. It seems that the architect had not specified sufficient insulation in the flat ceiling to take care of the amount of heat the engineer specified. The contractor had to put a hipped roof on top and put insulation on top of the old roof in order to provide sufficient comfort. In this case, the failure was the architect's and not the engineer's, but both had to pay for it with their insurance. So, be very careful that the architect designs a building in accordance with your calculations, or vice-versa.

In the downtown area of a city, a restaurant was installed on the first floor of a 20-story building with the kitchen exhaust blowing out the side wall. Unfortunately, the odors from the kitchen crossed the street, entered the City Hall, and caused considerable complaints. The result was that the city passed an ordinance requiring the exhaust to be run to the top of the building which, of course, put the restaurant out of business. This probably could have been solved more simply by using a high velocity fan in the exhaust, and by using a deflector louver to blow the exhaust air up vertically. Then odors would not have entered the adjoining 3-story building.

A restaurant in the center of a 1-million-population city had a roof exhaust that was spewing odors from the production of corned beef and dill pickles. The odors penetrated a jewelry store and a men's clothing store alongside.

Of course, this went to court and the judge selected me to try to settle the case. On investigation, I told the owner to quit making dill pickles because of the high vinegar content, to increase his exhaust over cooking areas, and to use an exhaust fan on the roof which blew the air vertically instead of horizontally. The restaurant owner followed these instructions for about 2 years and then went back to making dill pickles in the basement. I had to examine the operation, since he claimed he was not making them but had a truckload of pickles in the back yard. In his cooler, he had a barrel of vinegar with just an apron covering the top of it. He finally had to purchase the jewelry store and the clothing store alongside in order to stay in business.

We were hired once by an insurance company to look at a residence which had burned down and killed the occupant. The building was 90 years old, out in the country, and the fire marshal who accompanied us on this trip insisted he thought the deceased woman's son had killed her. On examination, it was found that all that was left was a basement and all the materials in the house had fallen into the basement. I asked the attorney to have the basement cleaned out and I would again examine it. When it was cleaned out, I went back in and found that there was an oil-fired furnace with an oval 250-gallon oil tank only about a foot away. In examining the furnace burner, which was burned white, I noticed that the connection for the oil tank was a flare fitting and had not been connected. The connecting tube from the oil tank was covered with soot which indicated that it had not been connected. It was then discovered that the son had disconnected the oil pipe, let the oil sink into the dirt floor of the house, and had set it on fire. He was convicted.

In another case, a man was standing on an aluminum ladder, painting the trim on a university dormitory window. He had two men on the ground holding his ladder. After he was through painting, he leaned hack, touching a 6,500-volt power line. The current went through his body, down the ladder, and killed the two men on the ground. It didn't even singe the painter's jacket. Again, the power company was sued and the families collected.

Check the statute of limitations in the state where you are designing equipment to see if it extends beyond 10 years. If you choose equipment, make sure it will last that long, or you may be sued. We had one case where there was no statute of limitations. An architect had designed a building in West Virginia. After 20 years, the brick work started to spall off the building. Since the contractor as well as the manufacturer of the brick had gone out of business, the architect was sued and the owner collected $100,000. This temporarily canceled out the architect's insurance.

Chapter 21

Architectural Provisions for HVAC

1. Provide sufficient space with equipment to allow for maintenance.

2. For combustion and refrigeration equipment, provide airtight wall between and with airtight doors, chemical sensors and alarms, and gas masks. Don't forget outside air intakes for combustion, and ventilation for refrigeration equipment.

3. On rooftop enclosures for cooling towers and air-cooled condensers, provide sufficient space for air intake and discharge, to prevent recirculation.

4. Make roof color light if possible.

5. Provide roof surface walkway material for access to equipment.

6. Place signs on doors to rooms used for mixed air to air conditioning units to prevent storage of chemicals or flammable materials.

7. Specify color coding of piping, with labels as to content and use.

8. Under electric contract, specify wiring arrangement to allow rearranging fixtures

9. Conform electric lighting and equipment heat in space to assure sufficient cooling capacity.

10. Provide electrical ground outside of building sufficient to prevent electrolysis in piping

11. Provide access to air conditioning units above ceilings.

12. Provide sufficient weight and mass for floors of penthouse- and roof-equipment rooms to prevent noise and vibration transmission.

13. Provide vapor seal on insulation on outside in the South, to prevent moisture entering; and on the inside in the North, to prevent moisture leaving the space, both on walls and ceilings.

14. If a security system is not specified, make provisions to connect to energy management system.

15. Provide space for energy management equipment in maintenance department.

16. Provide location for outside-air intakes to prevent contamination from cooling towers, toilet and plumbing vents, and external sources such as auto exhaust.

17. Undercut doors to toilet rooms to allow exhaust to bring air in.

18. Under plumbing, provide sufficient water pressure to supply at least 25 psi to highest point of HVAC system. Provide access to main drain lines from air conditioning unit drain pans for cleaning.

19. If energy management system is considered, have architect contact owner if remote-control alarms and security are desired, such as at home or by cellular telephone.

20. Provide shades or window coverings for windows on the sunny sides.

21. Discuss with architect these conditions so that your design will provide for good capacity maintenance and reasonable operation.

CHAPTER 22

PREVENTING SERIOUS ACCIDENTS

Following are some of the unforeseen problems (some of which are discussed in detail in reference books) that could have been prevented by construction inspection and maintenance.

1.   Hospital

A hospital was supplied with two different electric utilities with an automatic switch-over and three motor generator sets, all of which failed during a storm, requiring that an operation be performed by flashlight. Solution—run the motor generator set continuously at low load for surgeries.

2.   Industrial Plant

A plant had an 11,000 volt distribution electrical system to 22 4,000-kW transformers. Lack of insulating oil in transmission lines and surrounding connections to the transformers caused 10 of them to explode. One missed me by 10 minutes. Solution—Replace main feeder line, insulate transformer-connection terminals, and provide surge arrester in case of high voltage during storms.

3.   Bank

A main switchgear panel exploded, putting the bank out of business. Cause—loose connections on wiring to panel. Solution—Replace panel with proper wiring connections tightly made.

4.   Country Club

A poor wiring connection at main service entrance is shown in Figure 22-1. Wire rated at 200 amps was hot carrying only six amperes. Potential was fire. Solution—replace with a junction block for wiring.

5.   Office Building

Power plant on roof was oil- and gas-fired. Oil leaked, did not shut off when switched to gas. Explosion occurred, bulging the steel walls of the boiler out 18." Solution—Provide time delay on gas firing after oil shut-off. Also provide light-weight roof over boiler room so explosion would go vertically instead of sideways, damaging building (see Figure. 22-2).

6.   Paper Mill

Turbine drive on 400'-long shaft controlled by non-electric speed controller. Power failure in boiler plant caused turbine to over-speed, breaking up shaft and throwing pulleys a distance of three blocks. Solution—Providing an electrically operated speed control limit.

7.   Paper Mill

Paper breakage filling up basement area with heavy sheets 20' wide. Solution—Electronic control on staging speed of rolls in proportion to length and moisture content.

8.   Warehouse

Scuppers in parapet walls surrounding flat roof not large enough to relieve sufficient water during extremely heavy rainstorm. The 100,000-square-foot of roof collapsed causing $150,000 damage to property stored inside. Roof was designed for 3" of water. Construction of steel framing was improper for this particular area as 4" of water was sufficient to collapse the roof.

9.   Space Heater

Vent from space heater used a dryer vent with a back-draft damper. Result—two people died.

Figure 22-1. Main electric feeder connection. 200 amp wire hot with only 6 amp currency.

Figure 22-2. Rooftop boiler room. Use light-weight roof to project explosions upward.

10. Office Building
Penthouse boiler room—no outside air intake for boilers, high carbon monoxide level. Solution—Reversed exhaust fans to provide combustion air.

11. Office Building
Gas boiler plant supplying absorption refrigeration system. Carbon monoxide level 15%. Solution—Provide gauge on gas line to prevent high contamination in flue gases. Gas pressure too high for type of burner.

12. Refrigeration Service
Supply of Freon or CFCs in refrigeration system in the room in which an open flame is present may cause phosgene which is deadly. Solution—Provide for ventilation when working on this type of equipment.

13. Apartment House
Garden hose used for supplying natural gas, connected ahead of meter and run up the stairway with potential for disaster. Solution—Remove hose.

14. Office
Air conditioning system does not function. Examination of air-cooled compressor indicates that relay to start compressor will not close because contacts are covered with fire ants. Fire ant nests around air conditioner. Solution—Kill fire ants.

15. Electrocution
The 11 cases I was involved with consisted primarily of contact with cranes, antennas, and bodies with high-voltage transmission lines. In the case of cranes, do not touch the outside of a crane if the boom is anywhere near a power line. The personnel inside with gloves and rubber-soled shoes are usually not affected, but persons leaning on the tread or the outside surface may be killed.

Use of an aluminum ladder with personnel holding the ladder at the ground are in danger if the person on the ladder is anywhere near a power transmission line.

16. Electrical Substation

If you have an open electrical substation, make sure it is secured from any animals. In one case, a cat climbed up and shorted out a 26,000-volt line which blew the insulators off of the substation of the power company four miles away.

These may give you an idea of some of the things that might happen with poor inspection and poor maintenance.

Chapter 23

Operation of Auxiliary Equipment

Auxiliary equipment in most buildings, if over 5 stories, includes booster pumps for the water system, fire pumps for the fire-protection sprinkler system, and emergency generators for at least elevators and emergency lighting.

On domestic water pumps, the usual system today is a variable-speed pumping system providing at least 25 pounds per square inch pressure at the roof line, or if there is a cooling tower, at the top of the cooling tower. We have experienced a number of cases where there was insufficient water pressure at the roof which interfered with the flow of condenser water to the air conditioning systems and reduced their efficiency. Another system which doesn't seem to be used as much any more is a storage tank on the roof. Water is pumped up to this tank and has sufficient capacity for approximately 15 to 30 minutes of normal use, then the pump can cycle back on again.

You might consider putting a recorder on the power to this pump to see if it adds to the demand rate on the building and see if a storage tank on the roof may be a possible solution to reducing the peak kW when the pump is running. We are using storage systems for the air conditioning system, why not use storage for the pumping system?

If you have a recreational area like a golf course with high electric costs for irrigation pumping, golf cart charging, or any other high electric use that may be operated at night or in an off-peak period, it may be advisable to use two electric meters—one for off-peak and one for day use.

Golf course watering systems usually consist of two 50 HP pumps costing about $25,000 per year to operate. Most greens keepers will not consider shutting off water during the day because they test for leaks and want to keep pressure on control valves. Why not try a 5 HP pump to keep pressure on line with a normal use electric meter, then switch to off-peak at night for general sprinkling? You may save $10,000 per year. My course saves $450 per month by recharging 100 golf carts at night.

Thermal storage for cooling by making ice or cold water storage may reduce demand costs also.

If the weather is not too humid, fountains and waterfalls evaporate some water cooling off the area. Test the temperature. You may be able to use this through your refrigeration or air conditioning system to get better efficiency.

If you are near the ocean or a contaminated river, using a heat exchanger may be cheaper than a cooling tower.

Fire protection pumps normally have small auxiliary pump called a "jockey pump" which keeps pressure on the system so that when a fire occurs, the main pump comes on and water is immediately available at the sprinkler head. In addition to the sprinkler system, it is now recommended to have a smoke removal system in which outside air is pressurized to exit areas like stairways, elevator shafts, and so-called secure areas on each floor. These should be connected to the emergency system, but the air intakes to the fans should be provided with back-draft dampers so that air does not enter the building through this system when the air conditioning system is shut down. Open smoke removal systems also may provide a stack effect to remove heat from the building, creating a negative pressure.

If you have underground or enclosed garages, you do need ventilation from the standpoint of keeping the carbon monoxide level down. Carbon monoxide detectors operating the exhaust fans will save operating costs if the concentration of carbon monoxide is below health levels.

Parking Lot Lighting

If you have extensive parking lot space which may be quite a distance from the building, it may be a good idea to zone the lighting so that the part of the lot farthest from the building may be shut off first when most of the space is vacated, keeping the nearest part of the lot lit for security reasons. This will require investigation of the wiring system for the lighting (see Figure 23-1).

Waste Heat

Investigate the possibilities of using the waste heat from air conditioning systems, from boiler flues, and any other exhaust

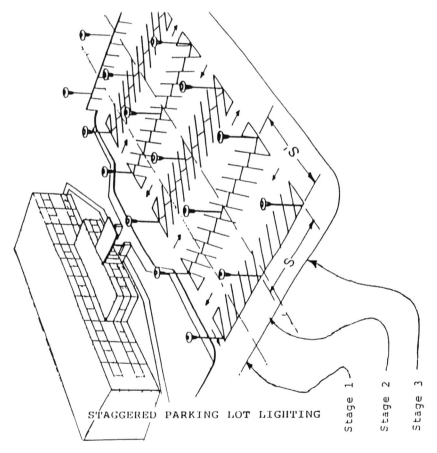

STAGGERED PARKING LOT LIGHTING

Stage 1 Stage 2 Stage 3

Figure 23-1.

device which is throwing heat away. There is equipment available that can take the heat from an air conditioning condensing system and heat water up to about 140° if desired. We used this device on a residence with a 40-ton air-cooled chiller which was using a 75 kW electric boiler for re-heat purposes, heating hot water. By the use of the heat from the air conditioning system, we were able to eliminate the use of the boiler. Since boiler flue gases are quite high, water can be heated in economizers on stacks if properly installed and if they do not interfere with the draft on the boiler. It is a good idea to use storage tanks for this water since this equipment may not be functioning continuously and you can gain as much as possible while it operates. Even though the water may not be heated to the desired temperature, it will at least bring most of the water up to a reasonable temperature so the additional heat will be considerably less than if you had to start from the initial temperature.

Heat pumps are really a form of generating heat from waste by which we take air or water and cool it down and throw it away and use that heat to heat the building. We have also used them for swimming pools using well water.

CHAPTER 24

Actual Equipment Installations

1. Cooling tower core—See Fig. 24.1

2. Thermostats in air conditioning equipment room—See Fig. 24

3. Remote thermostat control—See Fig. 24.3

4. Air-cooled condenser, putting hot air in outside intake—See Fig. 24.4

5. Baseboard radiation control—See Fig. 24.5

6. Indoor swimming pool air supply—See Fig. 24.6

7. Air conditioning unit mounting on sloping roof—See Fig. 24.7

8. Multiple ducts in joist space—See fig. 24.8

9. High mist discharge from cooling tower—See Fig. 24.9

10. In-line water meter—See Fig. 24.10

11. Diesel oil strainer for generator—See Fig. 24.11

12. Multiple grounding connection—See Fig. 24.12

Figure 24-1. Typical cooling tower core will plug up without water treatment.

Figure 24-2. Thermostats in air conditioning equipment room.

Figure 24-3. Thermostat sensor on mirror column; no adjustment (left). Thermostat controllers in separate room. Adjustment and temperature reading (right).

Figure 24-4. Air-cooled condenser heating up outside air intake.

Figure 24-5. Baseboard heaters. Do not modulate, end sections cool first.

Figure 24-6. Indoor swimming pool. Air blows toward windows.

Figure 24-7. Air conditioning units on sloping roof.

Figure 24-8. Ducts in joist space.

Figure 24.9 Mist discharge from cooling tower.

Figure 24-10. In-line water meter.

Figure 24-11. Diesel oil feed to generator. Install gauge on suction and discharge of pump, thermometer on line from oil tank.

Figure 24-12. Multiple grounding connection rod into conductive soil.

Chapter 25

Summary

In summary, it is suggested that you run your first run diagnostic tests if you have a problem with energy, air pollution, or health. If you cannot find the cause, check with the engineer who designed the system and see if he knows what may be wrong with the operation. If he cannot find it, have him get your maintenance contractor to test the equipment to see if it is functioning near its original design. If all that fails, then get an environmental engineer to run chemical and contaminant tests in the building.

If, after solving any obvious problems and considering all possible causes of illness, you still have personnel who appear to be suffering from their work environment, the only option is to send the person or persons to an allergist to see if there are any other environment elements that may be causing the difficulty.

Index